ACTIVE FILTER DESIGN

MODERN FRONTIERS IN APPLIED SCIENCE

Noise Performance Factors
in Communications Systems
by W. W. Mumford and E. H. Scheibe

Laser Applications
by Dr. William V. Smith

Microwave Engineer's Handbook
by Ted Saad

Logarithmic Video Amplifiers
by R. S. Hughes

Parallel Coupled Lines and Directional Couplers
by Dr. Leo Young

Microwave Filters Using Parallel Coupled Lines
by Dr. Leo Young

Microwave Semiconductor Control Components
by Dr. Kenneth E. Mortenson & Jose M. Borrego

Phased-Array Antennas
by Dr. A. A. Oliner & G. H. Knittel

Data Modem Selection and Evaluation Guide
by V. V. Vilips

Infrared To Millimeter Wavelengt[illegible]ors
by Dr. Frank R. Arams

Adaptive Electronics
by Dr. Wolfgang W. Gaertner

Gallium Arsenide Microwave Bulk
and Transit-Time Devices
by Dr. Lester F. Eastman

Variable Capacitance Diodes
by Kenneth E. Mortenson

Avalanche Transit-Time Devices
by Dr. George I. Haddad

Microwave Filters for Communication Systems
by C. M. Kudsia & M. V. O'Donovan

Radar Detection and Tracking Systems
by Dr. S. A. Hovanessian

Significant Phased Array Papers
by Dr. Robert C. Hansen

Stripline Circuit Design
by Harlan Howe

Microwave Diode Control Devices
by Robert V. Garver

Spectrum Analyzer Theory and Applications
by Morris Engelson & Fred Telewski

Bubble Domain Memory Devices
by Dr. Alan B. Smith

Ferrite Control Components
by Lawrence R. Whicker
(2 Volumes)
Vol. 1. Junction Circulators, YIG Filters and Limiters
2. Ferrite Phasers and Ferrite MIC-Components

Advances in Computer Communications
by Dr. Wesley W. Chu

Electronic Information Processing
by Dr. William V. Smith

The Entrepreneur's Handbook
by Joseph R. Mancuso
(2 Volumes)

Radars
by David K. Barton
(6 Volumes)
Vol. 1. Monopulse Radar
2. Radar Equation
3. Pulse Compression
4. Radar Resolution and Multipath Effects
5. Radar Clutter
6. Frequency Agility and Diversity

Microwave Integrated Circuits
by Dr. Jeffrey Frey

to my wife ELLEN

and children

HOWARD, BONNIE, and ROBIN

ARTECH HOUSE, INC.
affiliated with Horizon House – Microwave, Inc.
610 Washington Street
Dedham, Massachusetts 02026

PREFACE

This book is directed toward the average electronics design engineer to provide a direct method for the design of active filters using simple formulas and design tables. Although advanced mathematics and specialized knowledge were required to produce these tables, sophisticated designs can now be generated using basic rules and guidelines.

Tables of normalized active Elliptic-Function Low-Pass filters are provided in the Appendix. Using these tables this extremely powerful filter type can be easily designed.

A chapter has been included on predicting the properties of the tabulated designs in both the frequency and time domains. This section aids the design engineer in selecting the optimum filtering function.

Design of almost any filter type is possible using the precalculated data provided. By simple transformations, High-Pass, Band-Pass and Band-Reject filters can be designed from the tabulated Low-Pass values.

After selecting a suitable numerical design, the design engineer must insure that the physical realization is satisfactory for production. Practical problems have been anticipated. The effects of component tolerances, amplifier limitations and resistor and capacitor selection are discussed and guidelines are established which result in satisfactory filters. In the event any component adjustment is necessary, a number of alignment techniques are provided.

A chapter on uncoventional filter types is given. This includes selective notch filters, fixed and adjustable phase shift networks and delay equalizers.

Many topics are presented for the first time in any book and provide a novel solution to filtering problems. Formerly, Active Band-Pass filter design involved manipulation of unwieldy equations and complicated RC synthesis techniques. A design method is outlined that involves simple formulas in conjunction with the numerical tables and results in practical filters which are in agreement with theoretical predictions.

I would like to thank A. Bodony for his encouragement and assistance. I also wish to express my appreciation to A. Scott and L. Torre for their valuable help in preparation of the manuscript and to L. G. Burnell, B. Lurie and B. Norvell who were instrumental during my early years of designing filters.

Arthur B. Williams

TABLE OF CONTENTS

Chapter 1 Introduction to Active Filters

1.1 A Brief History of Active Filters 1
1.2 Some Misconceptions about Active Filters . 2
1.3 Frequency and Impedance Scaling 3
1.4 Using this Handbook 6

Chapter 2 Selection of Response Functions

2.1 The Complex Frequency Plane 9
2.2 Frequency Response Criteria 11
a. Low-Pass Filters 12
b. High-Pass Filters 12
c. Band-Pass Filters 15
d. Band-Reject Filters 19
2.3 Transient Response Considerations 19
a. Group Delay 19
b. Group Delay of Low-Pass Filters 20
c. Group Delay of Band-Pass Filters 20
d. Step Response 22
e. Step Response of Low-Pass and Band-Pass Filters 22
2.4 Butterworth Polynomials 22
2.5 The Chebyshev Approximation 24
2.6 The Maximally Flat Delay Characteristic . . 29
2.7 The Elliptic Function Response 39

Chapter 3 Design of Active Filters Using Tables

3.1 All-Pole Low-Pass and High-Pass Filters . . . 43
a. Low-Pass Filter Design 44
b. High-Pass Filter Design 48
3.2 Elliptic Function Low-Pass and High-Pass Filters 49
a. Low-Pass Filter Design 56
b. High-Pass Filter Design 60
3.3 Wide-Band Band-Pass Filters 67
3.4 Wide-Band Band-Reject Filters 69

Chapter 4 Selective Band-Pass Filters

4.1 Single Pole-Pair Realizations 75

a. Single Amplifier Realization 75
b. State-Variable Pole-Pair Configuration . . 77
4.2 An All-Pole Band-Pass Filter Design Procedure 80
a. Real-Pole Transformation 80
b. Complex Pole Transformation 81
4.3 Alignment Techniques 83
a. Frequency Adjustment 85
b. Q Adjustment 88

Chapter 5 Unconventional Active Filters

5.1 Notch Networks 89
5.2 Delay Equalizers and Phase-Shift Networks . 90
a. Adjustable Delay and Amplitude Equalizer 93
b. All-Pass Delay Equalizer 94
c. Adjustable Phase-Shift Network 100
5.3 State-Variable Elliptic Function Filters . . .100
a. State-Variable Building Block 101
b. Low-Pass Design 104
c. High-Pass Filters 106
d. Band-Pass Filters 109
5.4 Active All-Pass Delay Lines 114

Chapter 6 Realizability and Practical Considerations

6.1 Sensitivity to Tolerances 123
6.2 Amplifier Limitations 124
a. Open-Loop Gain and Phase Shift 126
b. Input and Output Impedance 128
c. Dynamic Range 129
d. Amplifier Selection 129
6.3 Passive Component Selection 131
a. Resistor Selection 131
b. Capacitor Selection 135

Chapter 7 Appendix

7.1 Tables of Normalized All-Pole Low-Pass Filters 140

Table 7.1a — Butterworth Normalized Low-Pass Values 140
Table 7.1b — .1db Ripple Chebyshev Normalized Low-Pass Values 141
Table 7.1c — .5db Ripple Chebyshev Normalized Low-Pass Values 142
Table 7.1d — Maximally Flat Delay Normalized Low-Pass Values 143
7.2 Table of Normalized Elliptic Function Low-Pass Filters 144
7.3 Tables of Filter Polynomial Roots 174
Table 7.3a — Butterworth Normalized Polynomial Roots 174
Table 7.3b — .1db Chebyshev Normalized Polynomial Roots 175
Table 7.3c — .5db Chebyshev Normalized Polynomial Roots 176
Table 7.3d — Maximally Flat Delay Normalized Polynomial Roots 177
7.4 Review of Operational Amplifier Fundamentals 178
7.5 References 186

Index 187

CHAPTER 1

Introduction to Active Filters

1.1 A BRIEF HISTORY OF ACTIVE FILTERS

Active Filters were first introduced in the thirties. However, early configurations were very cumbersone since Vacuum tubes were used as the gain elements. Bulky size, high power requirements and cost offered no match against passive LC Filters. In effect, Active Filters were born prior to the technology necessary for their support.

When the transistor was invented in the late forties, many of the earlier shortcomings were eliminated therefore revitalizing new interest in Active Filters. An early constraint was the need to minimize the number of active components. Gain was expensive so a cost consideration rule of thumb involved substitution of a transistor for an inductor on an equal basis. Unfortunately, minimizing the number of transistors restricted early Active Filters to simple Low-Selectivity circuits.

As transistors became more economical and their parameters improved, a design approach developed where transistors were configured into Operational Amplifiers having nearly infinite gain, very high input impedance and near zero output impedance. RC Feedback Networks were introduced around these amplifiers resulting in high performance Filters. In the Fifties, Modular transistorized Operational Amplifiers were commercially available. However, their high cost restricted Active Filters to applications where size rather than economy was the primary consideration.

The most influential factor affecting Active Filters was the introduction of the Fairchild UA709 Monolithic Operational Amplifier in the mid sixties. This was the first integrated circuit amplifier offering equal performance to its discrete counterpart and high yield in volume production.

Active Filters in the range of DC to 500KHz can now be readily built, using standard Low-Cost off-the-shelf Micro-electronic Operational Amplifiers. They are being widely used in many areas such as Aerospace, Telephone and Data Communications, Audio Systems, Precision Instrumentation, etc. This book is an attempt to provide straight-forward methods for the design of such Filters.

The Active Filters discussed will be limited to inductorless networks consisting of lumped resistors and capacitors interconnected with operational amplifiers. Although the use of discrete parts is intended, these circuits can be built using state of the art Micro-Electronic technology involving deposited R-C networks and monolithic operational amplifier chips. Laser resistor trimming can then be used where filter adjustment is required.

1.2 SOME MISCONCEPTIONS ABOUT ACTIVE FILTERS

Design of both Passive and Active Filters has generally been considered to be a form of "Black Magic" and has been restricted to the specialist. In all fairness, design of Passive Filters does require specialized knowledge not generally available from Engineering Schools or from Literature, such as coil winding techniques, special circuit transformations to produce practical component values, etc. In contrast, most Active Filters can be assembled using off-the-shelf components and can be designed in such a manner as to result in practical component values.

The function of a filter is to produce an Output to Input Frequency Response. This response can be expressed as a ratio of two polynomials in S where $S=J\omega$ and is referred to as a Transfer Function. The roots of the numerator polynomial are called the Zeros and the roots of the denominator polynomial are called the Poles. A desired Transfer Function can be provided by either a Passive or Active Filter. There is absolutely nothing different about the response of the Active Filter as compared to the Passive Filter (except for a constant Gain Multiplier) when both Filters have the same Poles and Zeros and are operating in their linear region.

In the past, Active Filters have had a stigma of instability. The tendency for oscillations was a result of mainly two factors; the use of relatively unstable circuit configurations and marginally stable amplifiers. The configurations provided in this handbook have been selected for their high stability. Use of standard Integrated Circuit Operational Amplifiers as directed assures amplifier stability.

A major disadvantage of Active Filters has been the lack of adjustment capability, thus placing the Frequency Response at the mercy of component tolerances whereas passive filters have tunable Inductors. The design techniques presented in this book indicate convenient methods for the adjustment of resonances where required such as for Narrow-Band Band-Pass Filters.

1.3 FREQUENCY AND IMPEDANCE SCALING

The frequency response of a given active filter can be shifted (or scaled) to a different region of the frequency axis by dividing either the resistor or capacitor values by a Frequency Scaling Factor (FSF). The FSF is the ratio of a reference frequency of the desired shifted response to the frequency of the given filter that has equivalent attenuation. Usually, 3db points are selected as reference frequencies for Low-Pass or High-Pass filters and the center frequency for Band-Pass filters.

Tables are provided in the appendix for Low-Pass filters having a 3db point at 1 radian. These filters are considered "Normalized" to 1 radian. To shift this 3db point to a desired frequency, the FSF must be determined. The FSF can be expressed as:

$$\text{FSF} = \frac{\text{Desired Reference Frequency}}{\text{Existing Reference Frequency}} \tag{1-1}$$

Both the numerator and denominator must be expressed in the same units, either Hertz or Radians where Radians equal $2\pi F$. The following example shows computation of the FSF.

Example 1-1

Desired: Equivalent response with 3db point of 1000Hz.

Given: Normalized Low-Pass Filter of Figure 1-1a and associated response shown in Figure 1-1b.

Result: Calculate FSF

$$\text{FSF} = \frac{2\pi 1000\text{Radians}}{1\ \text{Radian}} = 6280 \tag{1-1}$$

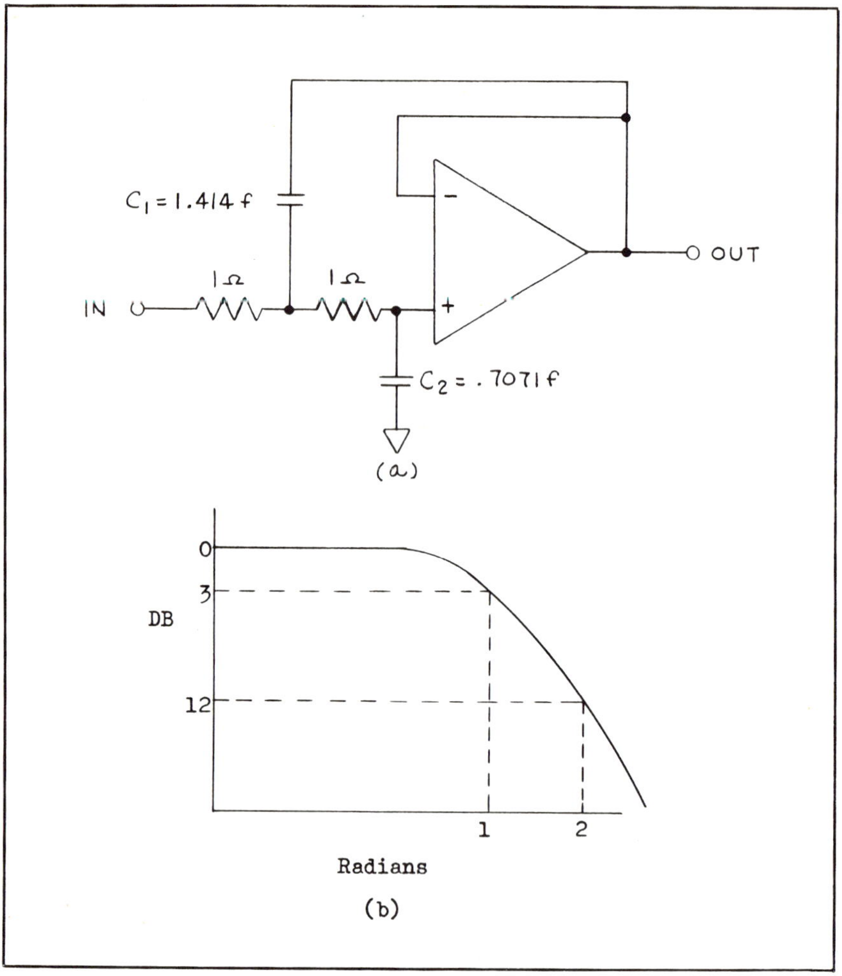

Figure 1-1
Normalized Low-Pass Filter and
Associated Frequency Response

Dividing the capacitor values by the FSF results in the denormalized filter of Figure 1.2a and the response shown in Figure 1.2b. Note that all points on the frequency axis have been multiplied by the FSF.

It is clear that the resulting capacitor values are too large to be practical. This is resolved by Impedance Scaling. A linear Active or Passive RC Network will retain its response if all the resistor values are multiplied by a factor Z and all the capacitors are divided by Z.

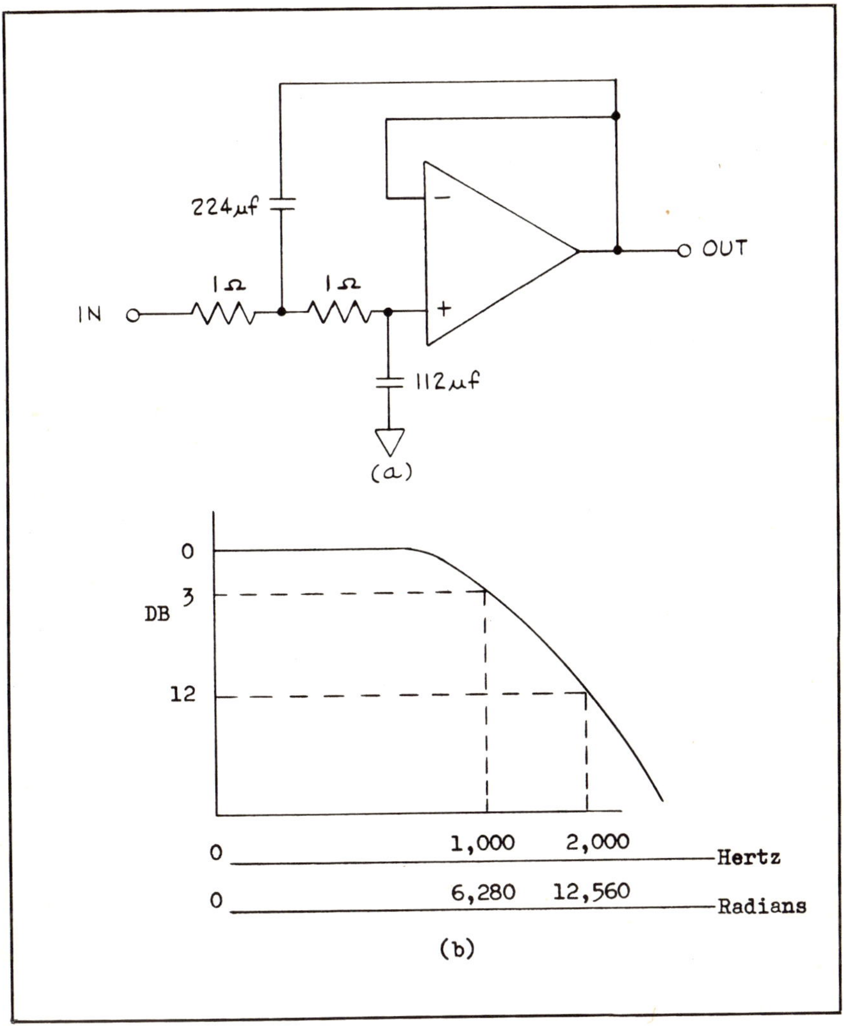

Figure 1-2
Denormalized Low-Pass Filter and
Associated Frequency Response

A basic example is a simple RC Network where the Time Constant is RC. If R is multiplied by Z and C is divided by Z, the Zs' cancel in the time constant expression.

In the circuit of Figure 1-2a, if the resistors are multiplied by a Z of 10,000 and the capacitors are divided by this Z, the circuit of Figure 1-3 results. The resulting values are certainly practical.

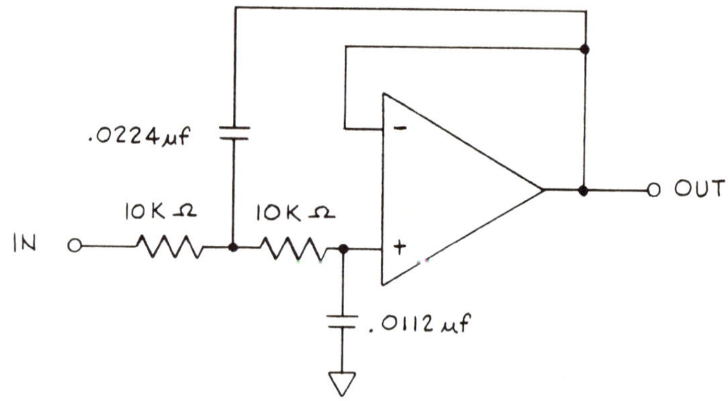

Figure 1-3
Denormalized Low-Pass Filter with Impedance Scaling

The process of Frequency and Impedance scaling can be combined into one step where:

$$C = \frac{C \text{ Normalized}}{ZxFSF} \tag{1-2}$$

$$R = R \text{ Normalized}xZ \tag{1-3}$$

1.4 USING THIS HANDBOOK

The process of designing Active Filters consists of two basic steps. Initially, a filter Transfer Function must be found that satisfies the frequency response specification. A tradeoff may be required between a filter type that offers a sharp rolloff but results in poor Transient Response and a less selective filter type that may require more sections to satisfy the specification but has less Transient Distortion. Chapter 2 discusses the various filter response functions and provides graphical data so that the design engineer can select the best filter type for his requirement.

After selection of a particular transfer function, a circuit must then be designed that realizes this response. Normalized tables are provided in the appendix for each associated transfer function described in Chapter 2. Use of these tables for Low-Pass, High-Pass and Wide-Band Band-Pass and Band-Reject Filters is described and demonstrated in Chapter 3. For Selective Band-Pass Filters Chapter 4 is applicable.

If the required filter is an unconventional type such as a Notch Network, a Delay Equalizer or Phase Shift Network or a State Variable Elliptic-Function filter, Chapter 5 presents a number of useful configurations and design procedures.

After selection of the transfer function and the subsequent design of the Filter, consideration must be given to the type and tolerance of the Resistors and Capacitors and to the type of Operational Amplifier to be used. Chapter 6 provides guidelines for optimum component selection.

Chapter 7 provides the normalized filter tables and an extensive list of references for further reading.

In addition, a section is included in Chapter 7 which reviews operational amplifier fundamentals for the reader who may not be familiar with their theory and application.

CHAPTER 2

Selection of Response Functions

2.1 THE COMPLEX FREQUENCY PLANE

A passive or active linear network has a frequency response that can be described by a Transfer Function. This Transfer Function can be expressed as

$$T(S) = \frac{E_{OUT}}{E_{IN}} = \frac{N(S)}{D(S)} \qquad (2\text{-}1)$$

N(S) and D(S) are numerator and denominator polynomials in S. If $J\omega$ were substituted for S, where $J = \sqrt{-1}$ and $\omega = 2\pi F$, the Transfer Function could then be directly evaluated.

Example 2-1

Given:

$$T(S) = \frac{1}{S^2 + 1.414S + 1} \qquad (2\text{-}2)$$

Verify that this transfer function is that of the circuit of figure 1-1.

Result:

a) Substitute $S = J\omega$

$$T(J\omega) = \frac{1}{1 - \omega^2 + J1.414\omega} \qquad (2\text{-}3)$$

b) Evaluate at ω = 0,1,2 Radians

ω	T(Jω)	Attenuation - DB
0	1	0
1	.707 $\angle$- 45°	-3
2	.242 $\angle$ - 137°	-12.3

This table is in agreement with the response of figure 1.1b.

The transfer function of example 2-1 is said to be a Second Order transfer function since the highest power of the denominator polynomial is two.

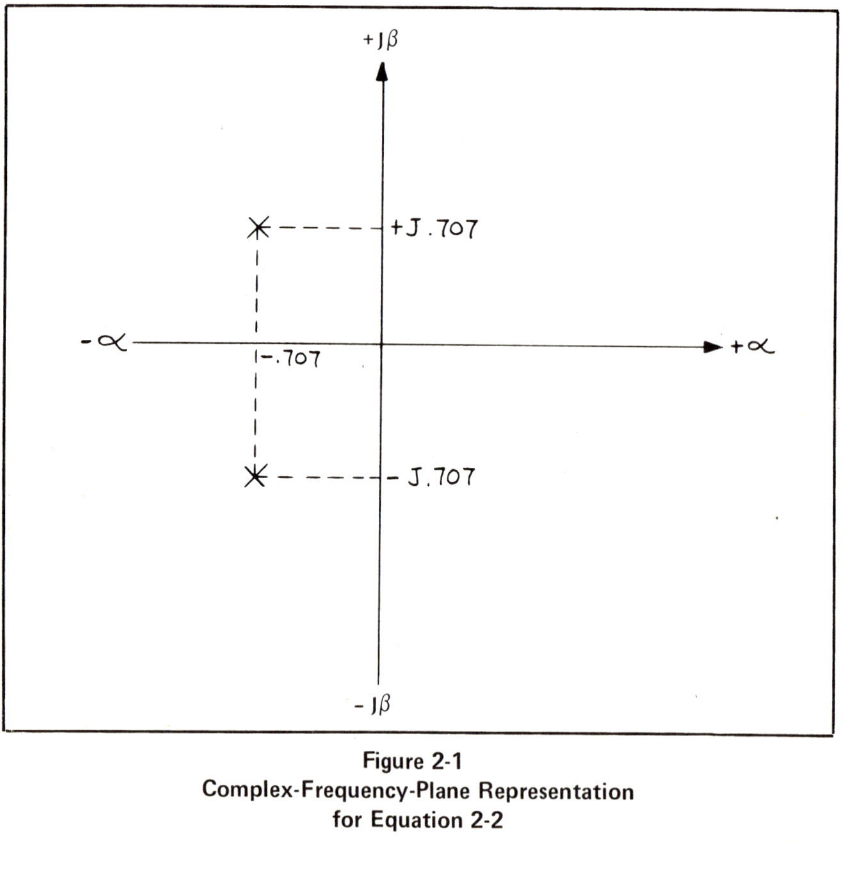

Figure 2-1
Complex-Frequency-Plane Representation for Equation 2-2

A transfer function can also be expressed in terms of the polynomial roots. The two roots of the denominator of equation 2-2 are:

$S = -.707 + J.707$ and

$S = -.707 - J.707$

These complex numbers can be represented in the Complex Frequency Plane. The abscissa is α, the real part of the root, and the Ordinate is β, the imaginary part. Numerator roots are shown as 0 and referred to as Zeros and denominator roots are shown as X and are referred to as Poles. Figure 2-1 illustrates the Complex Plane representation for example 2-1.

When the circuit of Figure 1-1 is denormalized to that of figure 1-2 the Transfer Function changes as well. The denormalized Transfer Function becomes:

$$T(S) = \frac{1}{S^2 + 8880S + 39.46x10^6} \qquad (2\text{-}4)$$

The denominator has the following roots:

$S = -4440 + J4440$ and

$S = -4440 - J4440$

These roots can be directly derived from the original normalized roots by multiplying the normalized co-ordinates by the FSF. Transforming a filter by scaling the component values by the FSF also scales the poles and Zeros (if any) by the FSF.

A mathematical restriction of Poles and Zeros is that they occur in pairs which are always conjugates of each other except for real axis poles and zeros which have no such restrictions and may occur singly.

2.2 FREQUENCY RESPONSE CRITERIA

A number of parameters define a Filters' characteristic. The most commonly specified requirement is Frequency Response. When given a Frequency response specification a filter type must be selected that satisfies this requirement. This is accomplished by transforming the specification to a normalized requirement and then comparing this normalized requirement to normalized curves in order to select a satisfactory filter. The chosen filter is then de-normalized to the required frequency.

These normalized curves consist of response curves of Low-Pass Filters having 3db of attenuation at 1 Radian except in the case of Elliptic-Function Filters where the attenuation at 1 Radian is equal to the Filter ripple. In order to use these curves, the filter requirement, which may be for a High-Pass, Band-Pass or Band-Reject type as well as Low-Pass must be converted into a normalized Low-Pass requirement. The following rules provide this conversion:

a. Low-Pass Filters

Suppose a filter is required that is 3db down at 1KHz and has at least 40db of attenuation at 4KHz. It is clear that a normalized Low-Pass Filter having 3db attenuation at 1 Radian and more than 40db at 4 Radians would meet this requirement when scaled by the proper FSF.

The general approach upon being given a cutoff frequency, F_c, and the frequency for the minimum required attenuation, F_s, is to form a ratio of the two which defines a Steepness Factor, A_s.

$$A_s = \frac{F_s}{F_c} \tag{2-5}$$

A normalized Low-Pass Filter which satisfies the minimum attenuation requirement at A_s Radians should be selected.

Example 2-2

Given: Low-Pass Filter Requirement
3db at 300Hz
50db minimum at 900Hz

$$A_s = \frac{F_s}{F_c} = \frac{900}{300} = 3 \tag{2-6}$$

Result: Normalized Low-Pass Filter requirement
3db at 1 Radian
50db minimum at 3 Radians

b. High-Pass Filters

Equation 2-2 is the Transfer-Function of a normalized Low-Pass Filter having the following frequency response:

$$T(S) = \frac{1}{S^2 + 1.414S + 1}$$

ω	DB
0	0
1	-3
2	-12.3

A Low-Pass to High-Pass Transformation of the normalized transfer function can be performed by substituting $\frac{1}{s}$ for S in Equation 2-2 with the following result:

$$T(S) = \frac{S^2}{S^2 + 1.414s + 1} \qquad (2\text{-}7)$$

ω	DB
∞	0
1	-3
.5	-12.3

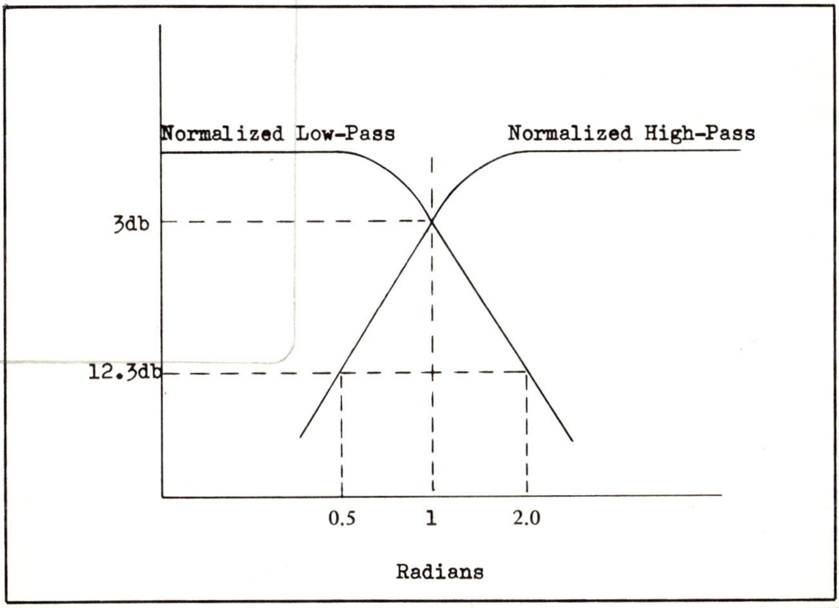

Figure 2-2
Low-Pass High-Pass Relationship

The response is clearly that of a High-Pass Filter. Observe that the Low-Pass attenuation values occur at High-Pass frequencies that are exactly the reciprocals of the Low-Pass frequencies. Figure 2-2 illustrates this relationship.

Assume that a High-Pass filter is required that is 3db down at 1KHz and has at least 40db of attenuation at 250Hz. If the 3db point were to be scaled to 1 Radian, then the 40db requirement would occur at .25 Radians. This normalized High-Pass requirement can now be transformed to a Normalized Low-Pass requirement by reciprocating the frequencies.

HIGH-PASS	NORMALIZED HIGH-PASS	EQUIVALENT NORMALIZED LOW-PASS
3db at 1KHz	3db at 1 Radian	3db at 1 Radian
40db min. at 250Hz	40db min. at .25 Radians	40db min. at 4 Radians

The selected normalized Low-Pass Filter can then be transformed into a Normalized High-Pass Filter and then scaled to a 1KHz cutoff by the FSF.

The intermediate normalized High-Pass transformation of the requirements can be bypassed by converting the High-Pass requirement directly to a steepness factor, A_s, and determining the normalized Low-Pass Filter that satisfies the minimum attenuation at A_s Radians. The A_s for High-Pass Filters is defined as the reciprocal of the A_s for Low-Pass Filters.

$$A_s = \frac{F_c}{F_s} \tag{2-8}$$

Example 2-3

Given: High-Pass Filter Requirement
3db at 300Hz
50db minimum at 100Hz

$$A_s = \frac{F_c}{F_s} = \frac{300}{100} = 3 \tag{2-9}$$

Result: Normalized Low-Pass Filter requirement
3db at 1 Radian
50db minimum at 3 Radians

c. Band-Pass Filters

Band-Pass Filters are generally classified into two categories, Narrow-Band and Wide-Band. A good rule of thumb is that if the ratio of the upper 3db Frequency to the Lower 3db Frequency is more than 1.5 the filter is considered Wide-Band.

Wide-Band Band-Pass Filter requirements can be separated into individual Low-Pass and High-Pass requirements which are satisfied independently. The resulting Low-Pass and High-Pass Filters are then cascaded to meet the overall specifications.

Example 2-4

Given: Band-Pass Filter requirement:
3db at 100Hz and 300Hz
40db minimum at 50Hz and 900Hz.

Result:

$$\frac{\text{Upper 3db Frequency}}{\text{Lower 3db Frequency}} = \frac{300\text{Hz}}{100\text{Hz}} = 3 \qquad (2\text{-}10)$$

Conclusion: Wide-Band type

High-Pass Requirement: 3db at 100Hz
40db minimum at
50Hz $A_s=2$

Low-Pass Requirement: 3db at 300Hz
40db minimum at
900Hz $A_s=3$

Narrow-Band Band-Pass Filters have a ratio of upper 3db Frequency to Lower 3db Frequency of less than 1.5. They cannot be treated as separate Low-Pass and High-Pass Filters.

We have seen earlier how a normalized Low-Pass Filter can be transformed into a High-Pass Filter by substitution of 1/S for S in the Transfer-Function. If S + 1/S were substituted for S in the Low-Pass Transfer-Function, a Band-Pass Filter would result centered about 1 Radian.

The Frequency response of the Low-Pass Filter is transformed into the bandwidth of the Band-Pass Filter at equivalent attenuation points. In other words if the Low-Pass Filter is Xdb at YHz (Radians), the Band-Pass Filter would have an Xdb <u>bandwidth</u> of YHz (Radians).

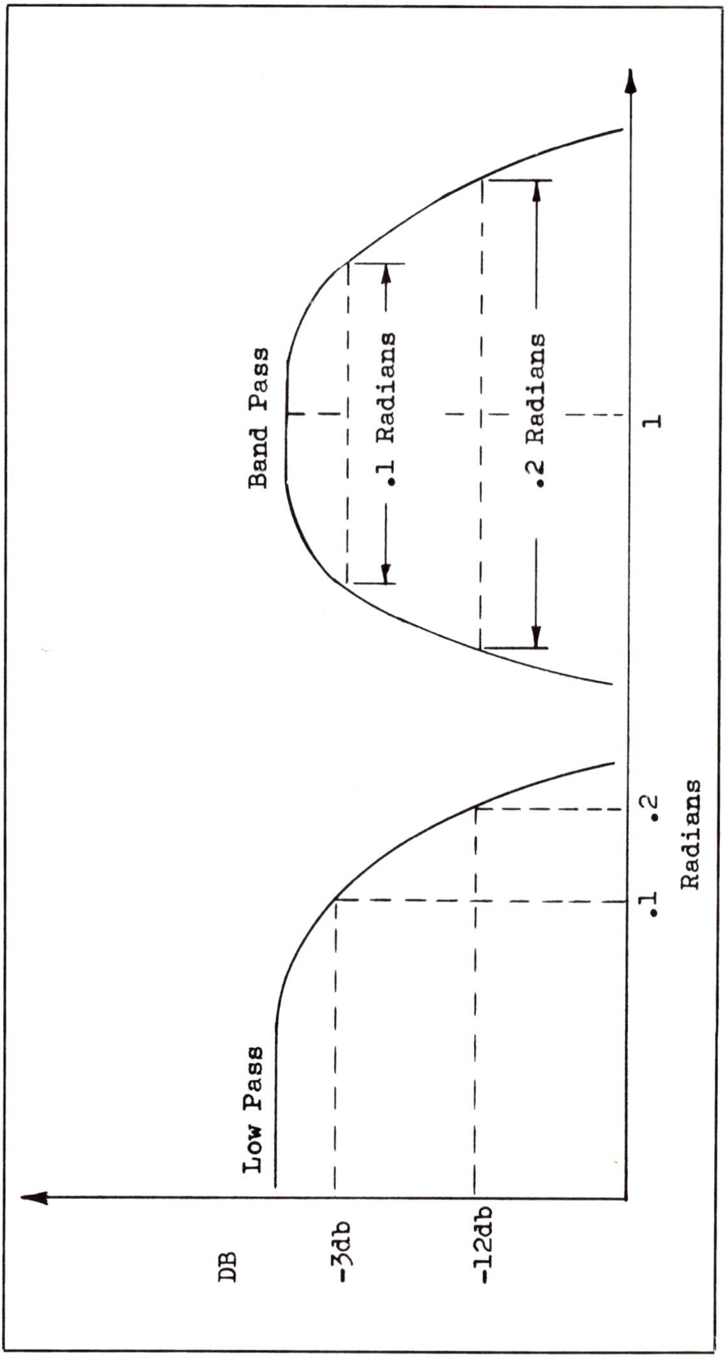

Figure 2-3
Low-Pass Band-Pass Relationship

Figure 2-3 illustrates this relationship where a Low-Pass Filter having a 3db cutoff of .1 Radians has a Low-Pass to Band-Pass transformation performed on its Transfer-Function resulting in a Band-Pass Filter have a center frequency of 1 Radian and a 3db Bandwidth of .1 Radians.

In order to use the normalized Low-Pass response curves provided in this handbook, a given Narrow-Band Band-Pass Filter requirement must be converted into a normalized Low-Pass requirement. This is performed in three stages as follows: 1) Convert the Band-Pass Filter specification into a response requirement for a Band-Pass Filter having Geometric Symmetry. Most Band-Pass Filter requirements are specified as arithmetically symmetrical i.e. symmetrical around the center frequency when plotted on a Linear-Linear axis. However, conventional[1] Band-Pass Filters are Geometrically Symmetrical i.e. symmetrical around the center frequency when plotted on a Linear-Log Axis with frequency on the Logarithmic axis.

To make the conversion, first calculate the Geometric Center Frequency F_0, using the following formula:

$$F_0 = \sqrt{F_1 \ F_2} \tag{2-11}$$

Where F_1 is the lower 3db frequency and F_2 is the upper 3db frequency. For very Narrow Band-Pass Filters where the ratio of F_2 divided by F_1 is less than 1.1 (Also expressed as less than 10% Bandwidth) the filter approaches Arithmetic-Symmetry so:

$$F_0 = \frac{F_1 + F_2}{2} \tag{2-12}$$

At equal attenuation points, corresponding frequencies on both sides of F_0 in the specification must be geometrically related to each other. This means that the two frequencies, F_A and F_B must satisfy the following relationship:

$$F_A \ F_B = F_0^2 \tag{2-13}$$

Modify the specification table by alternately calculating the corresponding geometric frequency for each stop-band frequency using formula 2-13. Two frequency pairs will result from each original pair. Discard the pair having the wider separation representing a less severe requirement.

[1] A special class of Band-Pass Filters are Arithmetically-Symmetrical but beyond the scope of this book.

2) All response requirements at the various attenuation levels must be converted to band-widths by subtracting the lower frequency from the upper.

A steepness factor, A_s, can now be defined as:

$$A_s = \frac{\text{Stop-Band Band-Width}}{\text{3db Band-Width}} \tag{2-14}$$

3) The normalized Low-Pass curves can now be entered using the A_s as the Radian Frequency. The selected Low-Pass Filter type can be transformed into the required Band-Pass Filter as described in Chapter 4.

Example 2-5

Given: Band-Pass Filter requirement
Center Frequency 1000Hz
3db Points ±150Hz (850Hz, 1150Hz)
40db Minimum ±300Hz (700Hz, 1300Hz)

Result: Calculate Geometric Center Frequency

$$F_o = \sqrt{F_A \; F_B} = \sqrt{850 \times 1150} = 989\text{Hz} \tag{2-15}$$

Compute two Stop-Band Geometric Frequency Pairs

Using F_A = 700Hz

$$F_B = \frac{F_o^{\,2}}{F_A} = \frac{(989)^2}{700} = 1397\text{Hz} \tag{2-16}$$

$$F_B - F_A = 697\text{Hz}$$

Using F_B = 1300Hz

$$F_A = \frac{F_o^{\,2}}{F_B} = \frac{(989)^2}{1300} = 752\text{Hz} \tag{2-17}$$

$$F_B - F_A = 548\text{Hz}$$

Modified Band-Pass Requirement
F_o of 989Hz
3db Band-Width 300Hz
40db Band-Width 548Hz

$$A_s = \frac{\text{40db Band-Width}}{\text{3db Band-Width}} = 1.83 \tag{2-18}$$

d. Band-Reject Filters

Band-Reject Filters having a ratio of 1.5 or more of the upper band-edge cutoff to the lower band-edge cutoff can be described as Wide-Band Band-Reject Filters. The filter specification can be separated into individual Low-Pass and High-Pass requirements. Filter selection proceeds as discussed in Sections 2.2a and 2.2b for the Low-Pass and High-Pass filters. Section 3.4 illustrates the complete design procedure including how the two filters can be combined (paralleled) to obtain the Band-Reject filter.

Another class of Band-Reject filters are Notch Networks. These filters are used to reject a single frequency or a narrow band of frequencies. Unlike the other filter types discussed, Notch Networks are designed directly at the frequencies of interest. Section 5.1 discusses selection of the design parameters and illustrates the design technique.

2.3 TRANSIENT RESPONSE CONSIDERATIONS

Frequency Response does not completely define the transmission properties of a filter. A very significant factor is the Transient Response of the circuit. Although filters are measured using sine waves, system applications involve applications of complex waveforms.

Any complex waveform can be represented by a number of sine waves at different frequencies using Fourier Analysis. In order to minimize distortion of the modulation represented by the complex waveform, each of the significant sine waves of the spectrum must not be displaced from each other in time. Thus each frequency component must undergo the same delay in passing through the filter network.

a. Group Delay

Delay to the information-containing sidebands of a modulated signal is called Group-Delay. It is defined in the following manner:

A filter exhibits a phase shift as a function of frequency which can be stated as $\Phi(\omega)$ where Φ is the phase shift in Radians at any frequency, ω.

Group-Delay is then defined as the following derivative:

$$Tgd = \frac{d\,\Phi(\omega)}{d\,\omega} \qquad (2\text{-}19)$$

Since Group-Delay is the derivative of the phase versus frequency response, for a constant Group-Delay, a Linear phase shift is required.

b. Group Delay of Low-Pass Filters

Group-Delay versus frequency curves for various Normalized Low-Pass Filters is presented in sections 2.4 through 2.6. These curves can be applied to Denormalized Low-Pass Filters by the following denormalization:

1) Divide the Delay Axis by $2\pi\ F_c$ where F_c is the filter 3db point in Hz.
2) Multiply the Frequency Axis by F_c.

The Delay variation over the band of interest can now be determined.

It should be mentioned that, generally, the absolute delay is not significant whereas the delay variation is of prime importance.

c. Group Delay of Band-Pass Filters

When a Low-Pass Filter is transformed to a Narrow-Band Band-Pass Filter, the shape of the Low-Pass delay curve is transformed to an almost symmetrical curve mirrored about the center frequency, F_0. However, the symmetry and Low-Pass shape is distorted, somewhat in proportion to the filter band-width. The Band-Pass filter delay curve can be approximated from the normalized curve as follows:

1) Divide the Delay Axis by πF_{bw} where F_{bw} is the 3db bandwidth in Hz.
2) Multiply the Frequency Axis by $F_{bw}/2$.
3) The curve can now be interpreted as the delay incurred for a frequency deviation of the frequency axis, from F_o.

Example 2-6

Given: Normalized Delay Curve of Figure 2-4a where the associated Normalized Low-Pass Filter is to be transformed to a Band-Pass Filter having an F_o of 1,000 Hz and a 3db Bandwidth of 20Hz.

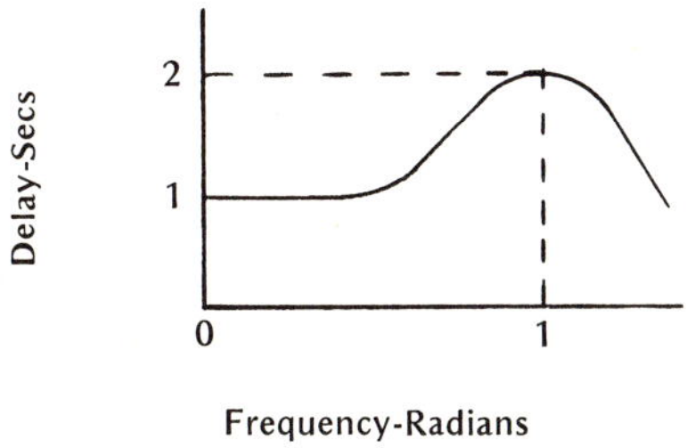

Figure 2-4a
Normalized Low-Pass Delay

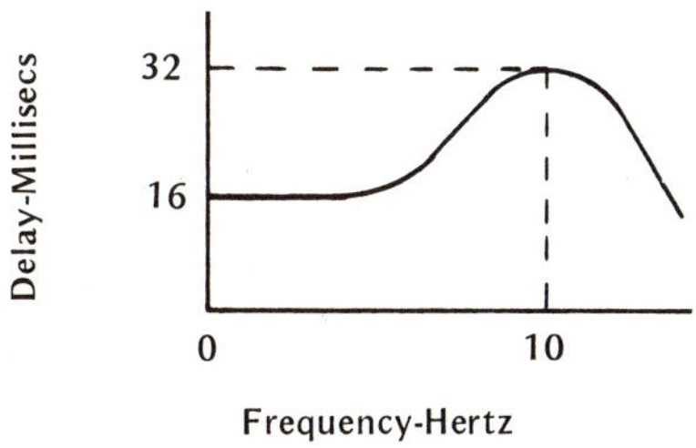

Figure 2-4b
Denormalized for Bandwidth of 20Hz

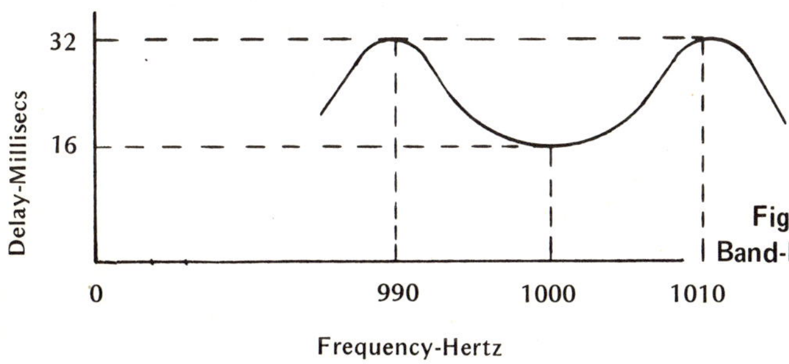

Figure 2-4c
Band-Pass Delay

Result: Dividing the Delay Axis by π F_{bw} yields:

$$\frac{1 \text{ sec}}{\pi\ 20} = 16 \text{ Milli-seconds}$$

$$\frac{2 \text{ secs}}{\pi\ 20} = 32 \text{ Milli-seconds}$$

Relabeling the frequency axis using $F_{bw}/2$ as the multiplication factor results in Figure 2-4b.

The curve can now be interpreted for the Band-Pass case, as shown in Figure 2-4c.

d. Step Response

Delay distortion does not directly indicate the extent of the distortion of the modulated signal. A more meaningful parameter might be the Step Response, especially in modulation systems where abrupt amplitude steps occur. Sections 2.4 thru 2.6 contain normalized Low-Pass Step Response curves consisting of the time response to a step of unity amplitude.

A useful criteria is the amount of step response overshoot. It is desireable that the amount of overshoot be minimized for accurate pulse reproduction. It is also desirable that the transient portion of the response die out rapidly.

e. Step Response of Low-Pass and Band-Pass Filters

The Step Response curves presented in Sections 2.4 thru 2.6 can be denormalized for Low-Pass filters by division of the time axis by $2\pi F_c$.

For Band-Pass Filters the time axis should be divided by $2\pi F_{bw}$. The curve then approximates the envelope of the AC response to a step of the center Frequency, F_o. Figure 2-5 illustrates the Low-Pass to Band-Pass Step Response transformation.

It should be mentioned that in the Low-Pass to High-Pass transformation the delay & step-response shapes are not preserved.

2.4 BUTTERWORTH POLYNOMIALS

As shown in section 2-1, filter responses can be described by Transfer Functions. By proper selection of these Transfer Functions the

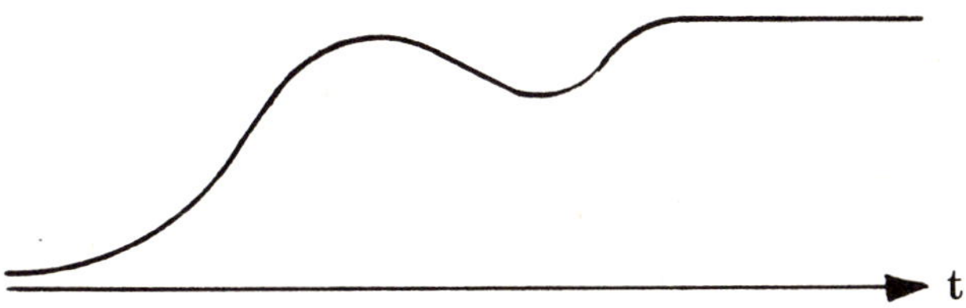

Figure 2-5a
Denormalized Low-Pass Step Response

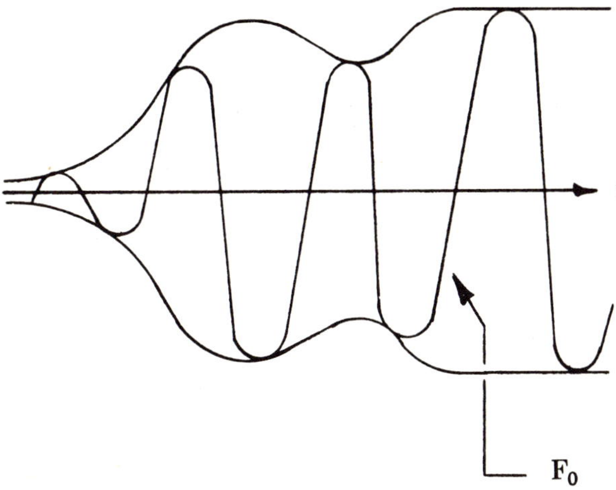

Figure 2-5b
Narrow-Band Band-Pass Step Response

filter response can take on various shapes. Sections 2.4 thru 2.7[1] discusses these different shapes which are a direct result of the polynomials of the Transfer Function Numerator and Denominator i.e. the Zeros and Poles respectively.

The Butterworth Low-Pass family of filters are All-Pole networks since the Butterworth Transfer Function has either unity or a constant multiplier in its' numerator indicating the absence of Zeros.

Normalized Butterworth polynomials have roots (poles) which all fall on a circle having a radius of unity, when plotted in the $J\omega$ plane.

Butterworth filters are probably the most popular filter type. The frequency response has the flattest possible response, at OHz for Low-Pass Filters, Center Frequency for Band-Pass Filters, and Infinite Frequency for High-Pass Filters. Beyond the 3db cutoff, the attenuation rate increases and eventually attains N x 6db/Octave, where N is the order of the polynomial, also referred to as the order of thc filtcr.

Butterworth filters have reasonably good phase response. They exhibit moderate steepness of attenuation and result in practical component values that are not critical. However, they have a rounded response in the vicinity of cutoff which might make then undesirable for some applications. Nevertheless they should be used wherever possible because of their favorable characteristics.

The Butterworth Low-Pass attenuation can be expressed as:

$$DB = 10 \, \text{Log}_{10} \, (1 + \omega^{2n}) \qquad (2\text{-}20)$$

Figure 2-6 presents a family of normalized attenuation curves up to N = 10.

Normalized Group Delay and transient Response are illustrated in figures 2-7 and 2-8 respectively.

2-5 THE CHEBYSHEV APPROXIMATION

Figure 2-9a illustrates the attenuation of an ideal Low-Pass filter. The desired response has no attenuation from DC to 1 Radian and infinite attenuation beyond cutoff.

The Butterworth approximation to this behavior is shown in Figure 2-9b. The approximation is relatively poor in the cutoff region while adequate at frequencies removed from cutoff.

A steeper rate of descent to the stop band and a more rectangular corner response is achieved using the Chebyshev approximation as

[1] Figures' 2-6 thru 2-8 and 2-10 thru 2-18 reprinted from Anatol I. Zverev, "*Handbook of Filter Synthesis*", John Wiley and Sons, Inc. 1967.

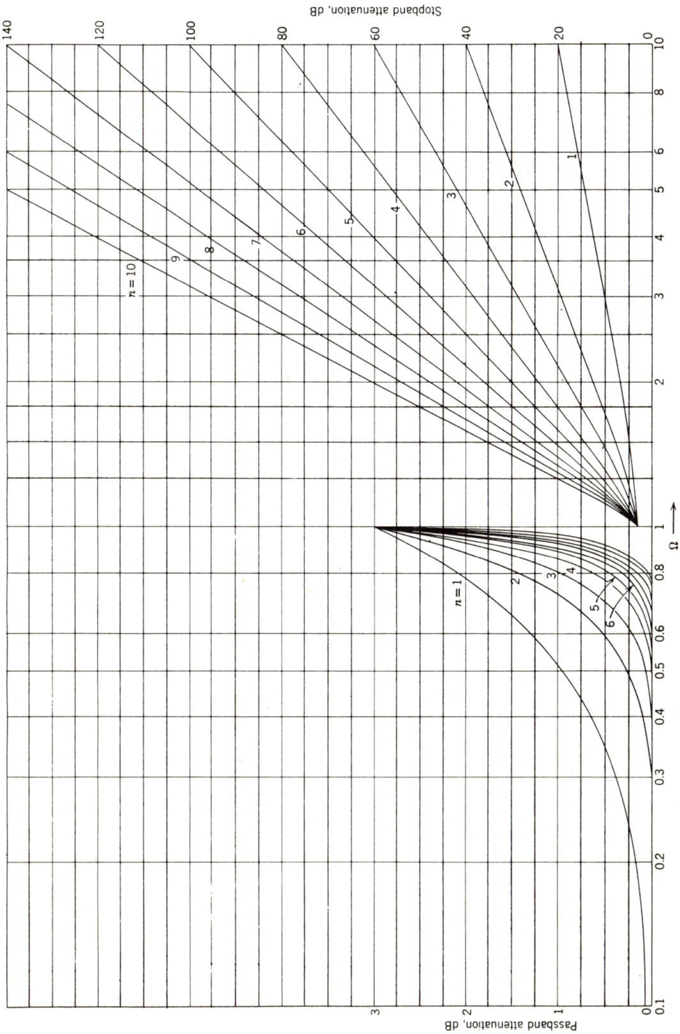

Figure 2-6 Attenuation characteristics for Butterworth filters.

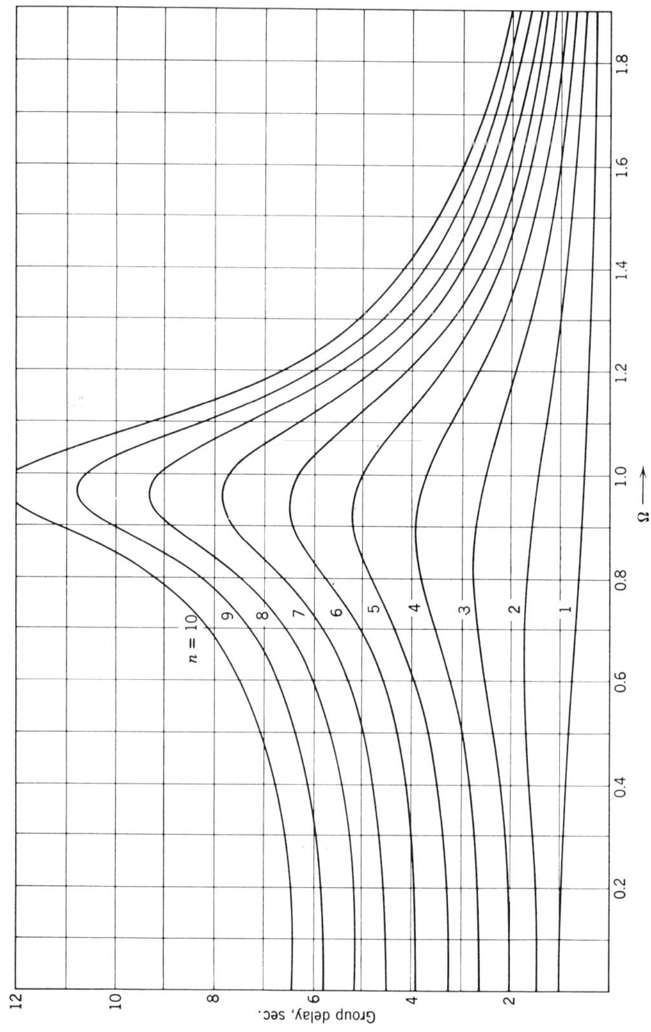

Figure 2-7 Group-delay characteristics for Butterworth filters.

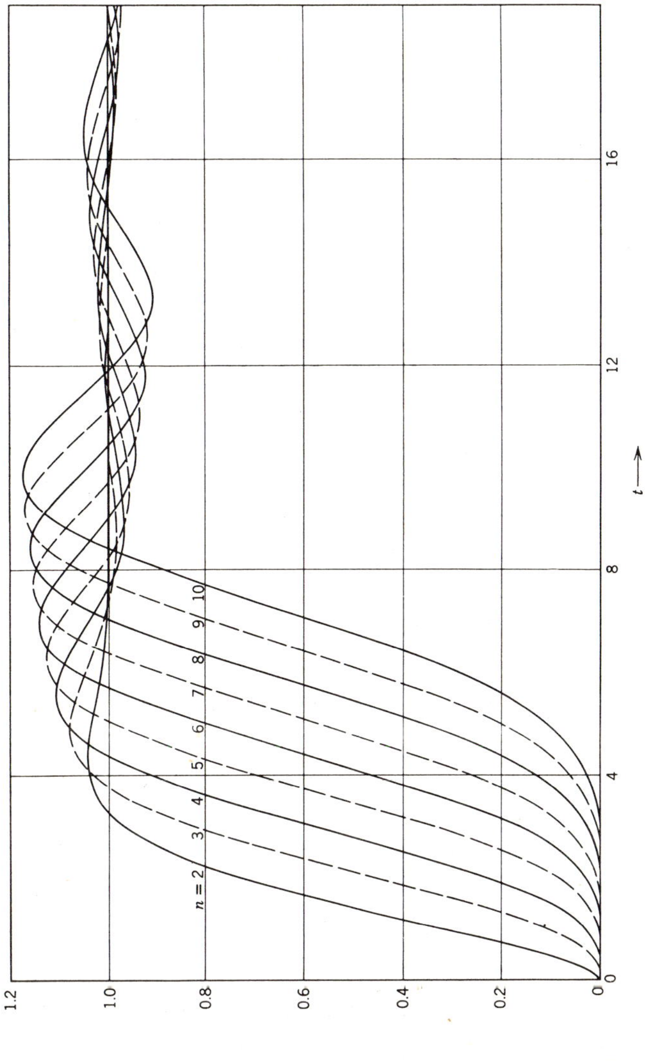

Figure 2-8 Step response for Butterworth filters.

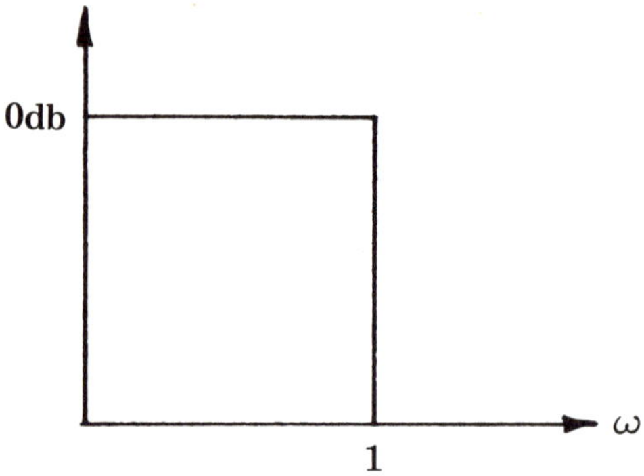

Figure 2-9a
Ideal Low-Pass Filter

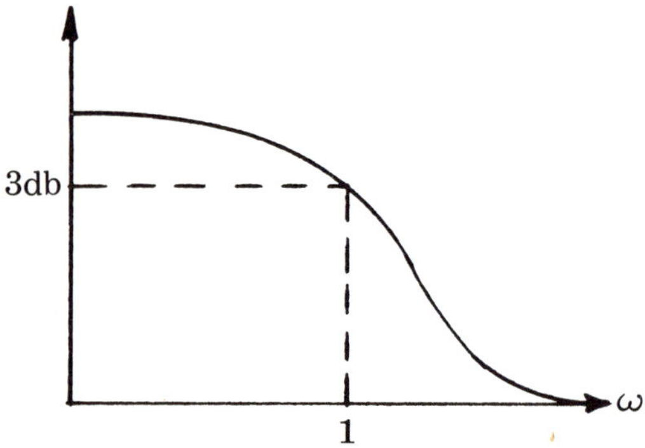

Figure 2-9b
Butterworth Low-Pass Filter

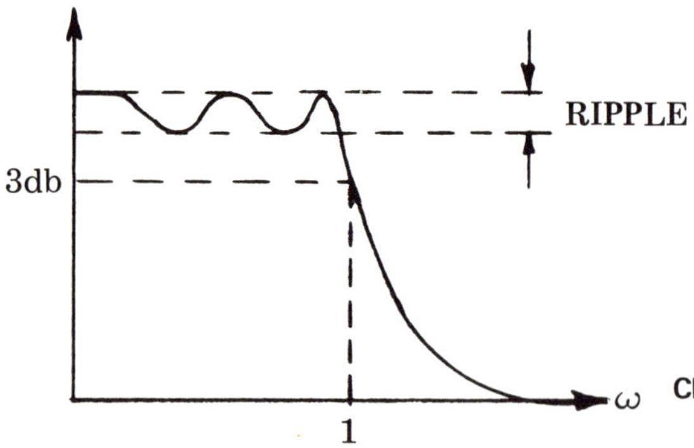

Figure 2-9c
Chebyshev Low-Pass Filter

shown in Figure 2-9c. This is achieved at the expense of allowing some ripple in the pass band.

For a given order of filter, the larger the ripple, the steeper the rate of descent to the stop band. However, the delay and transient behavior is poorer than the Butterworth family and could become intolerable for large ripples. The component values also become more critical and less practical with increased ripple.

Chebyshev filters, like the Butterworth, are also All-Pole filters. For a given order N, Chebyshev filters provide the steepest possible descent into the stop band of the All-Pole types.

While Butterworth Filters have their poles on a circle, the Chebyshev family of polynomials have poles on an ellipse.

Unlike the Butterworth family, Chebyshev filters slightly exceed N x 6db/Octave during the first octave beyond cutoff resulting in a steeper initial descent to the stop-band. However, all All-Pole filters eventually attain N x 6db/Octave at frequencies sufficiently removed from cutoff.

Figures' 2-10 thru 2-15 depict the Normalized Attenuation, Delay, and Step-Response for filters of .1db and .5db ripples.

2.6 THE MAXIMALLY FLAT DELAY CHARACTERISTIC

For pulsed systems, where faithful pulse reproduction is more important than attenuation, the maximally flat delay (Bessel) family of filters are useful. This family is similar to the commonly known Gaussian filters but provides a better approximation to a constant delay.

The attenuation slope near cutoff is very poor in comparison to Butterworth or Chebyshev filters. Even one or two octaves removed from cutoff the attenuation is not quite equivalent to the other types. However, the Delay is extremely flat especially for the higher order networks. The step response has no over-shoot.

Figures 2-16 thru 2-18 illustrate the attenuation, delay, and step response of this filter type.

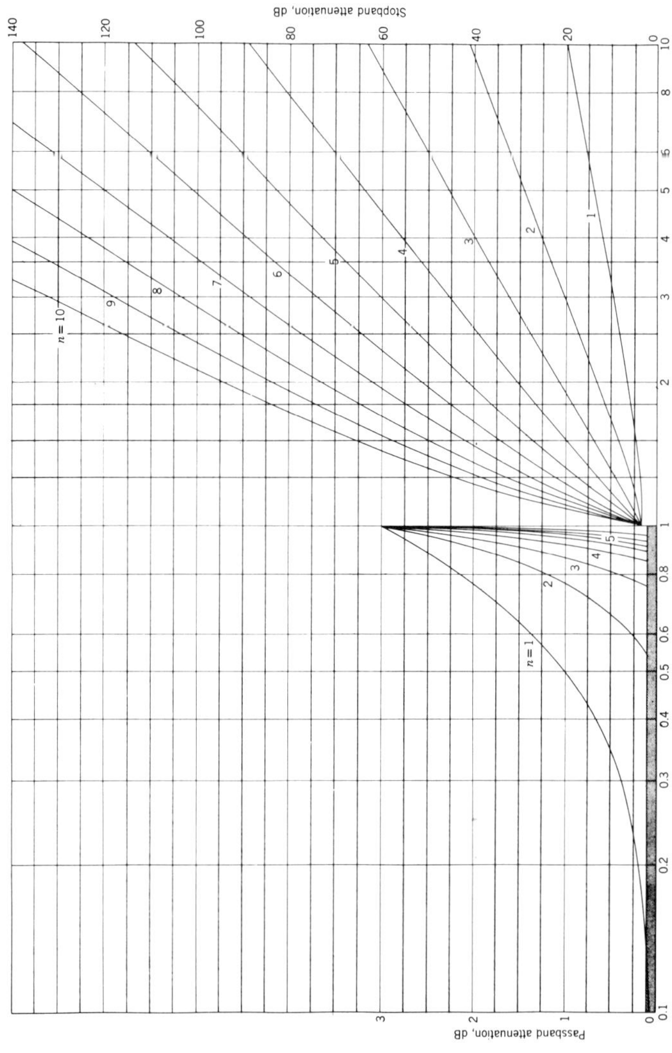

Figure 2-10 Attenuation characteristics for Chebyshev filter with 0.1 dB ripple.

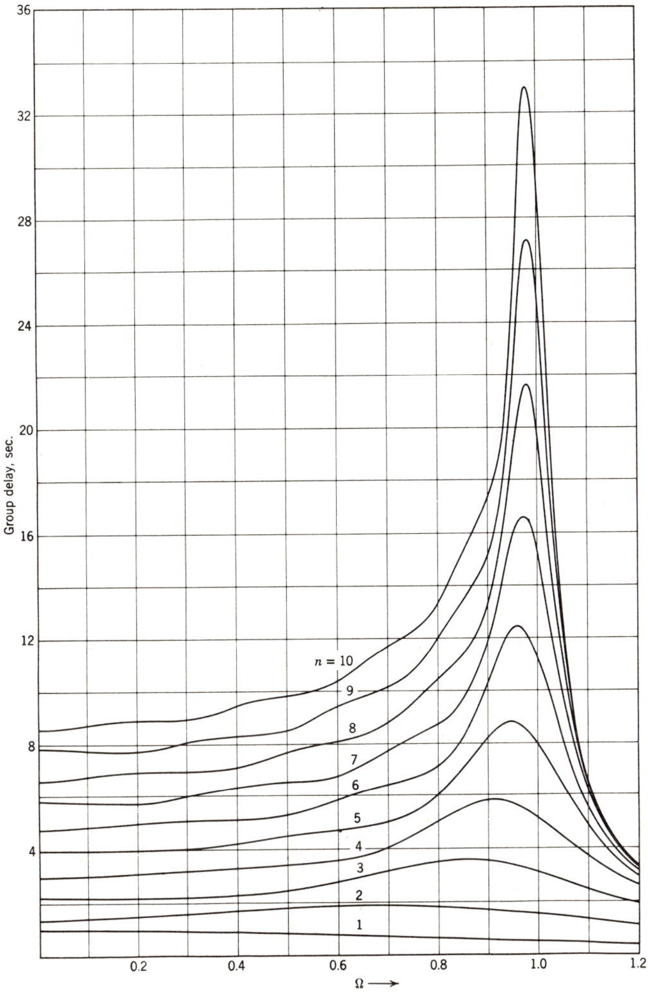

Figure 2-11 Group-delay characteristics for Chebyshev filter with 0.1 dB ripple.

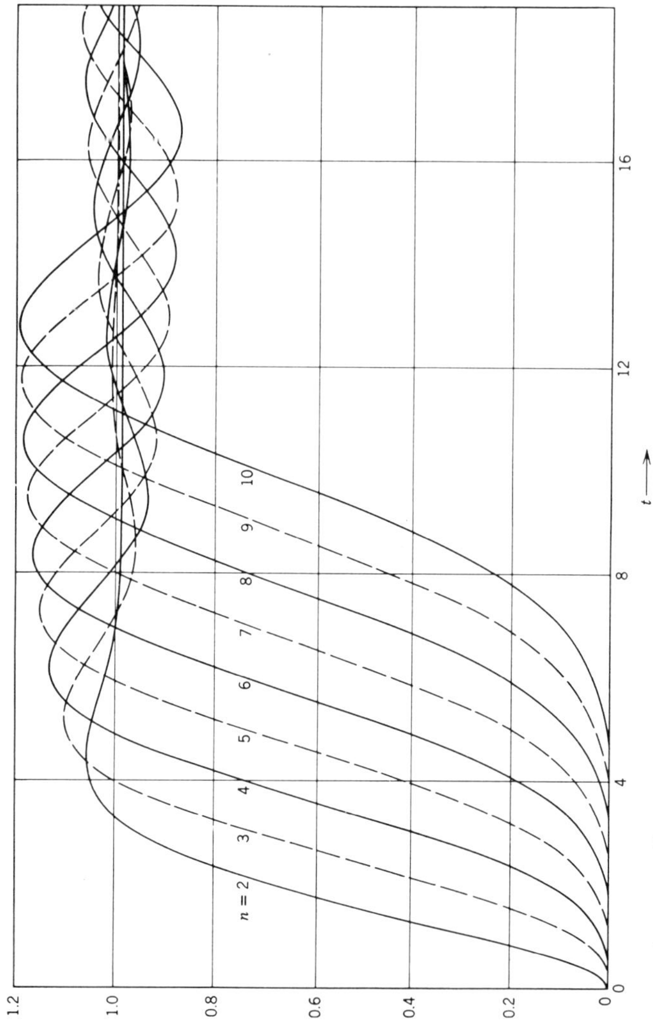

Figure 2-12 Step response for Chebyshev filters with 0.1 dB ripple.

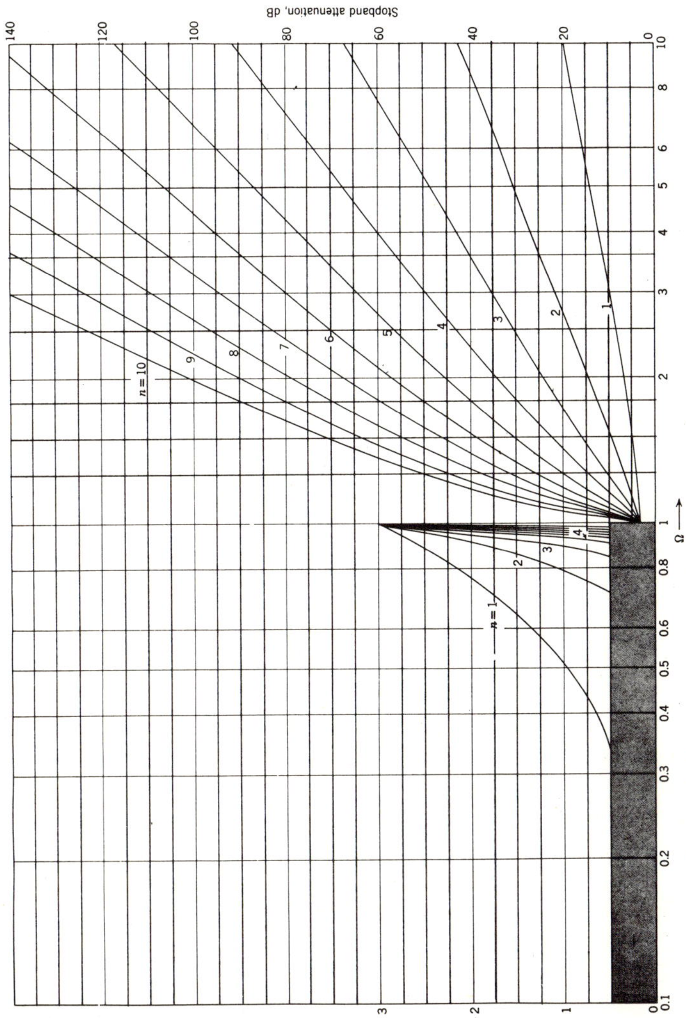

Figure 2-13 Attenuation characteristics for Chebyshev filter with 0.5 dB ripple.

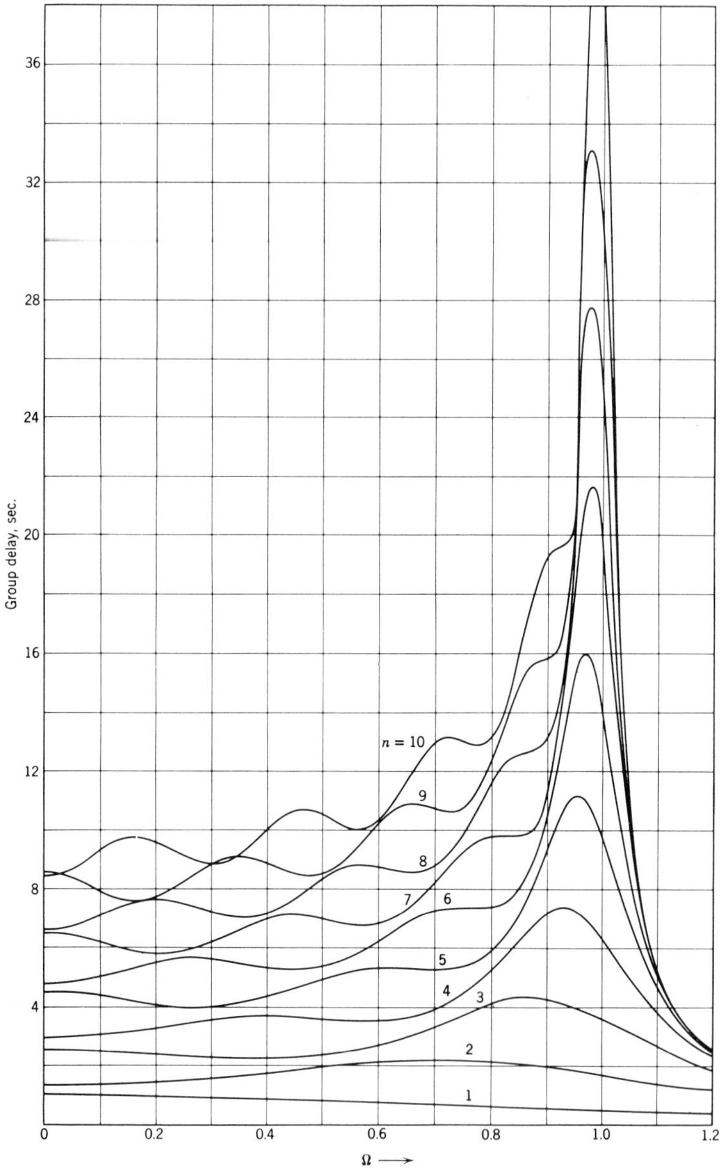

Figure 2-14 Group-delay characteristics for Chebyshev filter with 0.5 dB ripple.

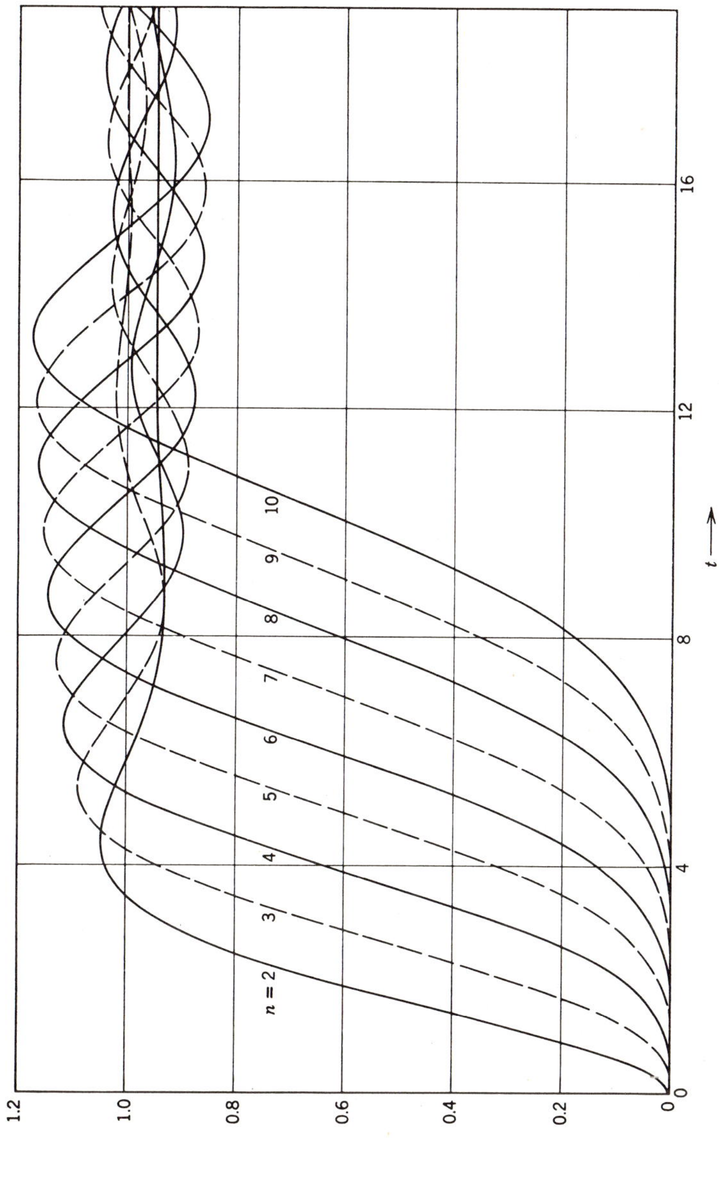

Figure 2-15 Step response for Chebyshev filters with 0.5 dB ripple.

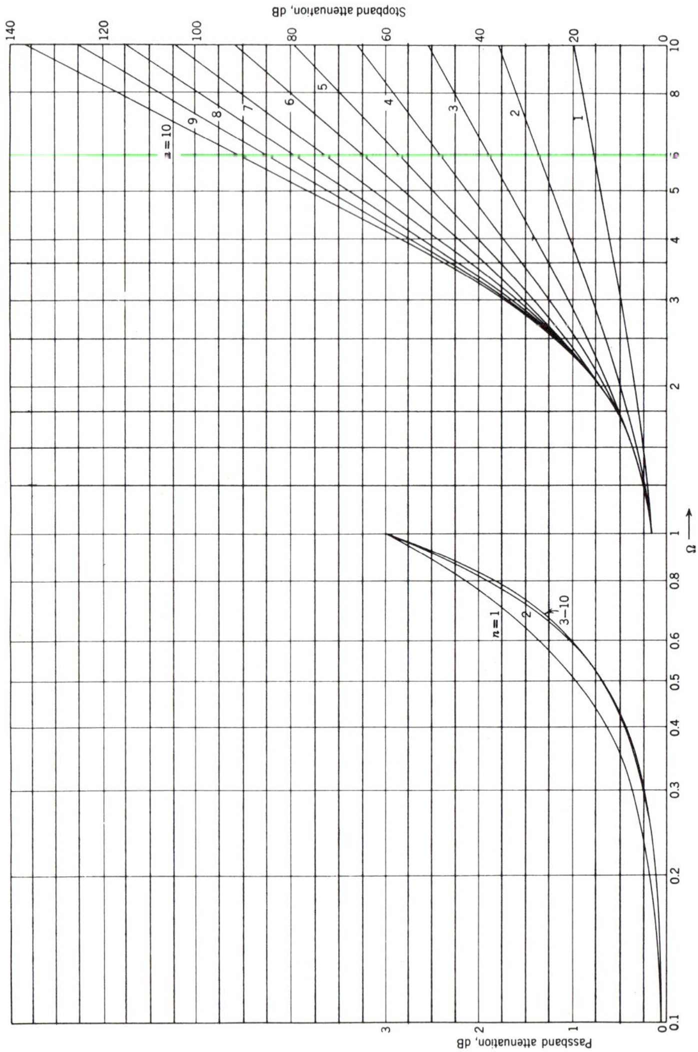

Figure 2-16 Attenuation characteristics for maximally flat delay (Bessel) filters.

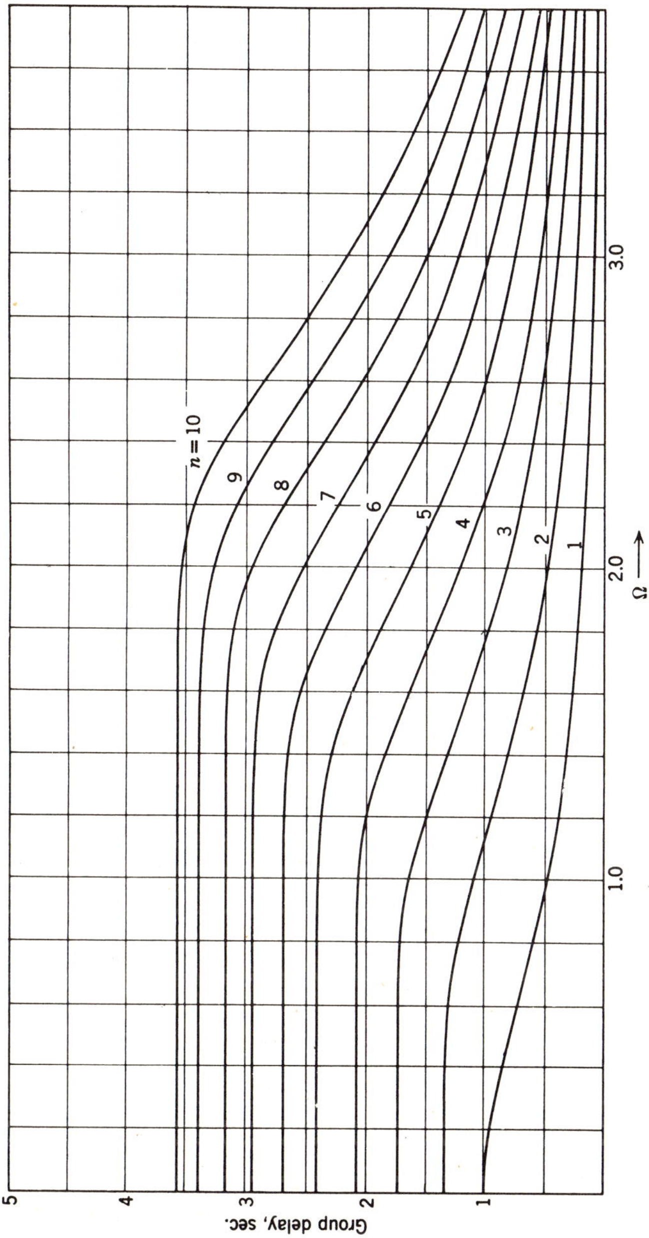

Figure 2-17 Group-delay characteristics for maximally flat delay (Bessel) filters.

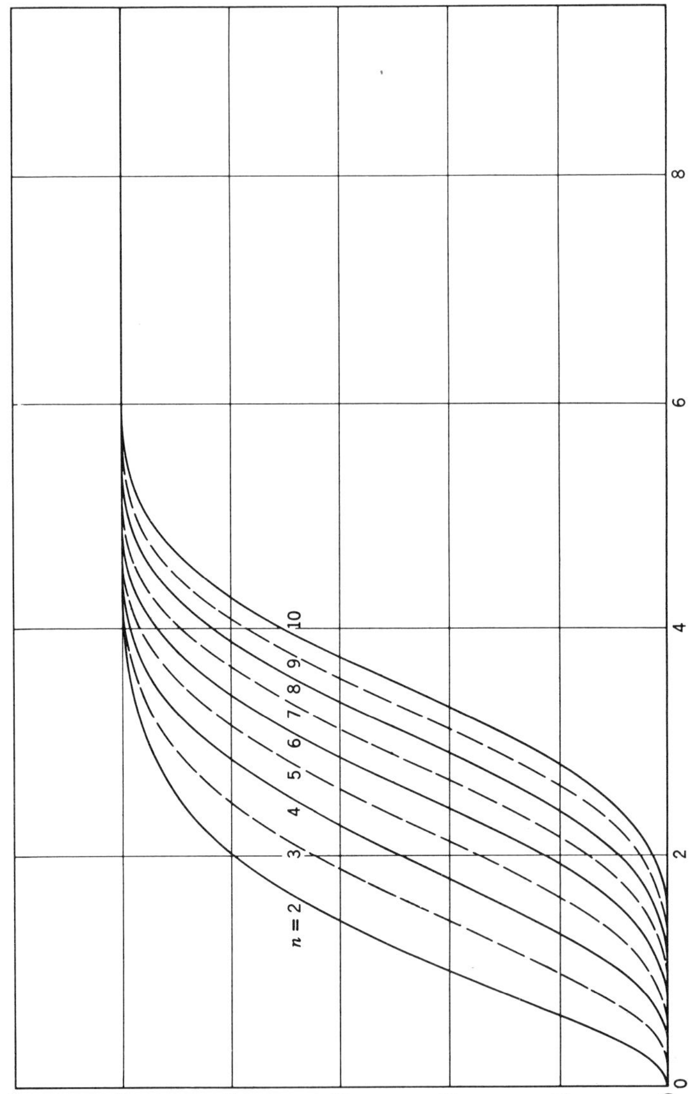

Figure 2-18 Step response for maximally flat delay (Bessel) filters.

2.7 THE ELLIPTIC FUNCTION RESPONSE

The Butterworth Low-Pass Filter is extremely flat at very low frequencies and somewhat rounded in the vicinity of cutoff. Its Transfer function is of the form 1/P(s) where P(s) is a polynomial in S having roots on a unit circle in the complex frequency plane.

The Chebyshev filter has a steeper transition from the passband-to stopband region but is not significantly sharper near cutoff compared to Butterworth filters. The Transfer Function is of the same form as the Butterworth but the denominator roots are on an ellipse rather than a circle.

Until now, we have discussed only All-Pole Filters. These filters have infinite rejection only at infinite or zero frequencies. The Elliptic Function filter contains zeros as well as poles at finite frequencies. These poles and zeros are positioned so that equal ripple passband behavior results, similar to Chebyshev filters. The return lobes are equal in the stop-band.

Using transmission zeros allows the steepest rate of descent possible for a fixed number of poles. Figure 2-19 illustrates a comparison between a five Pole Butterworth, a five Pole .1db ripple Chebyshev and a five Pole .1db ripple Elliptic-Function Low-Pass Filter having two finite zeros. Clearly the Elliptic Function Filter is far superior to the other types in the cutoff region.

Superior performance is achieved at the expense of return lobes and increased circuit complexity. Return lobes generally do not present any problems from a user point of view since a minimum stopband attenuation is required and a filter is selected whose return lobes meet this requirement. Generally speaking, the more the performance required from a filter, the more complex the realization and the more critical the component values. However, although each filter section is more critical and complex than Butterworth or Chebyshev sections, less sections are required for equivalent attenuation. For light filtering tasks, the Butterworth or Chebyshev types are preferable. For steep filter specifications the Elliptic Function Family is a must.

A number of terms unique to Elliptic Function filters must be defined. Figure 2-20 shows a normalized Elliptic Function Low-Pass Filter response. The following definitions apply:

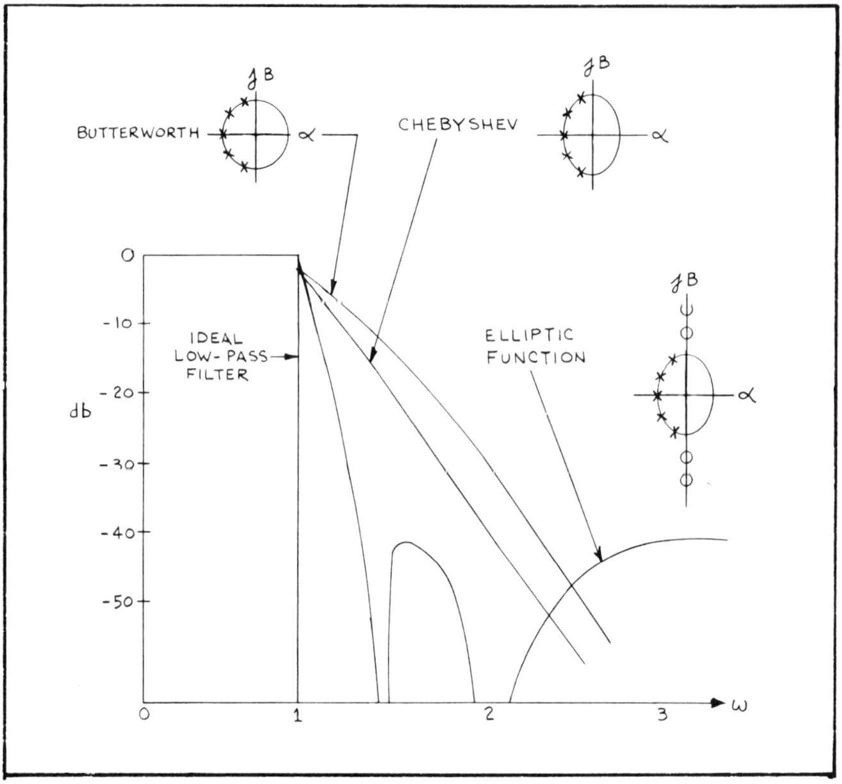

Figure 2-19
Comparison between Butterworth,
.1db Chebyshev and
.1db Elliptic Function Filter for N=5

R_{db} = Passband Ripple in DB
A_{db} = Minimum Stop-Band Attenuation in DB
ω_s = Lowest Stop-Band Frequency at which A_{db} occurs, in Radians.

The Elliptic-Function filters tabulated in the Appendix are categorized by order N, R_{db}, A_{db} and ω_s. Since the latter three parameters define the limits of the Pass-Band and Stop-Band response, normalized response curves are unnecessary.

Like the Chebyshev family of filters, delay variation and step response deteriorate with increasing order N and passband ripple R_{db}. Further deterioration occurs as the steepness is increased (reducing ω_s).

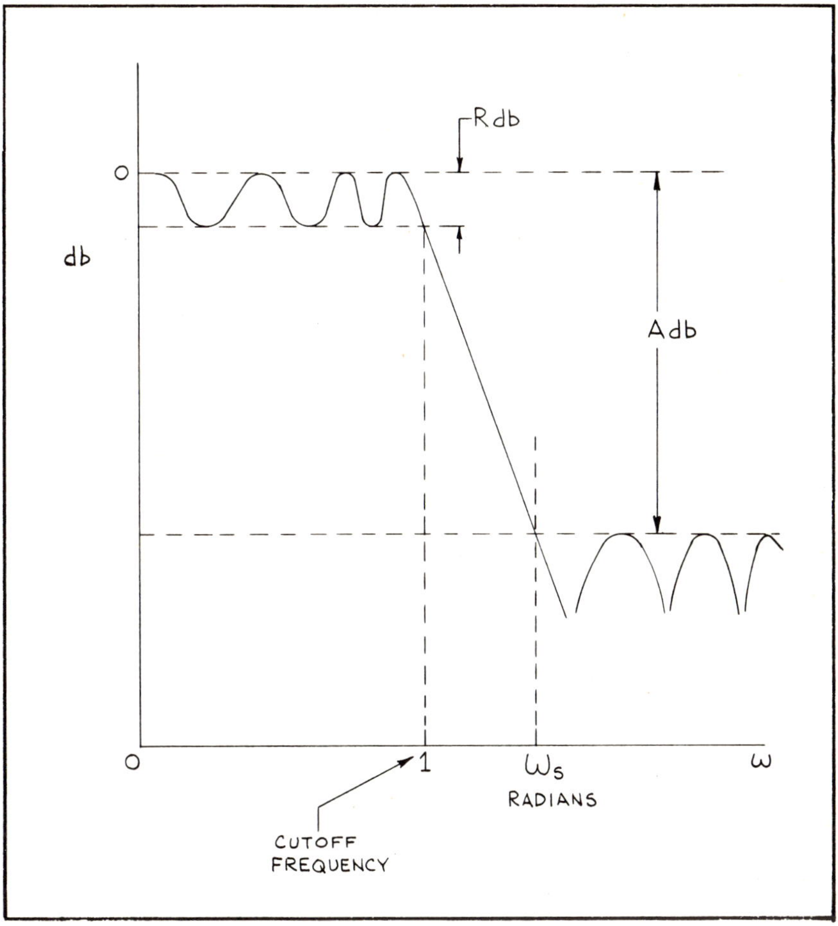

Figure 2-20
Normalized Elliptic Function
Low-Pass Response

CHAPTER 3

Design of Active Filters Using Tables

3.1 ALL-POLE LOWPASS AND HIGHPASS FILTERS

A passive RC network has poles that lie only on the negative real axis of the complex frequency plane. However the transfer functions discussed in Chapter 2 have complex poles. By the addition of an active element to the passive RC network, complex poles can be acheived (see Fig. 3-1).

For economy and stability a Voltage Follower configuration is used. This provides a gain of unity for an operational amplifier of high open-loop gain.

Two basic Low-Pass configurations are used for two-pole and three-pole realizations. The two-pole section acheives a pair of complex conjugate poles. The three-pole section produces a pair of complex conjugate poles and a single real axis pole. If the filter order N is an even number, N/2 two-pole sections are required. If N is odd, (N-3)/2 two-pole sections and a single three-pole section are necessary. This occurs because odd order filter transfer functions have only one real pole and the rest complex conjugate pairs. Even order filters have complex pole-pairs only.

Figures' 3-1a and 3-1b show the Low-Pass two-pole and three-pole sections. All resistor values are 1 ohm in the normalized filter. Capacitors C1, C2 and C3 are tabulated in the table 7.1[1] of the appendix. These values correspond to the normalized transfer functions of section 2 having a 3db cutoff of 1 radian.

[1] Tables 7.1a thru 7.1d reprinted from "Electronics", August 18, 1969, copyright McGraw-Hill, Inc., 1969.

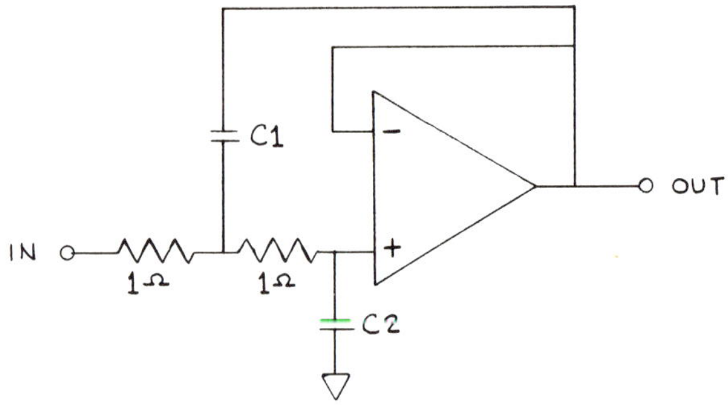

Figure 3-1a
Two-Pole Low-Pass Section

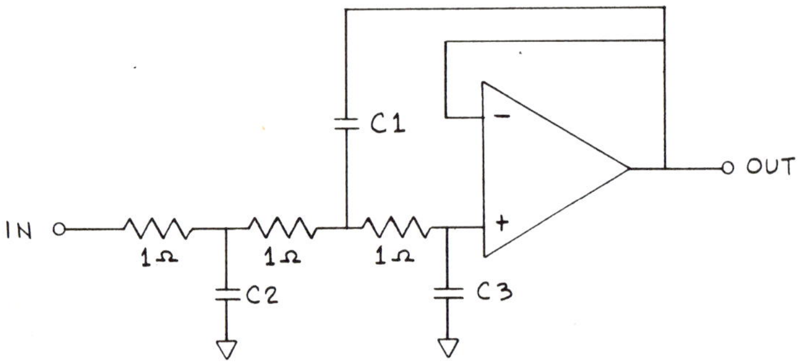

Figure 3-1b
Three-Pole Low-Pass Section

a. Low-Pass Filter Design

To design a Low-Pass Filter, a filter type must first be selected using the guidelines of section 2. The engineer then looks up the appropriate values from Table 7.1. It is helpful to sketch a schematic at this point showing the normalized cascaded sections.

The normalized filter is denormalized by dividing the capacitor values by Z x FSF where:

Z = Desired Impedance Scaling Factor

FSF = Frequency Scaling Factor ($2\pi F_c$ where F_c is the 3db cutoff frequency)

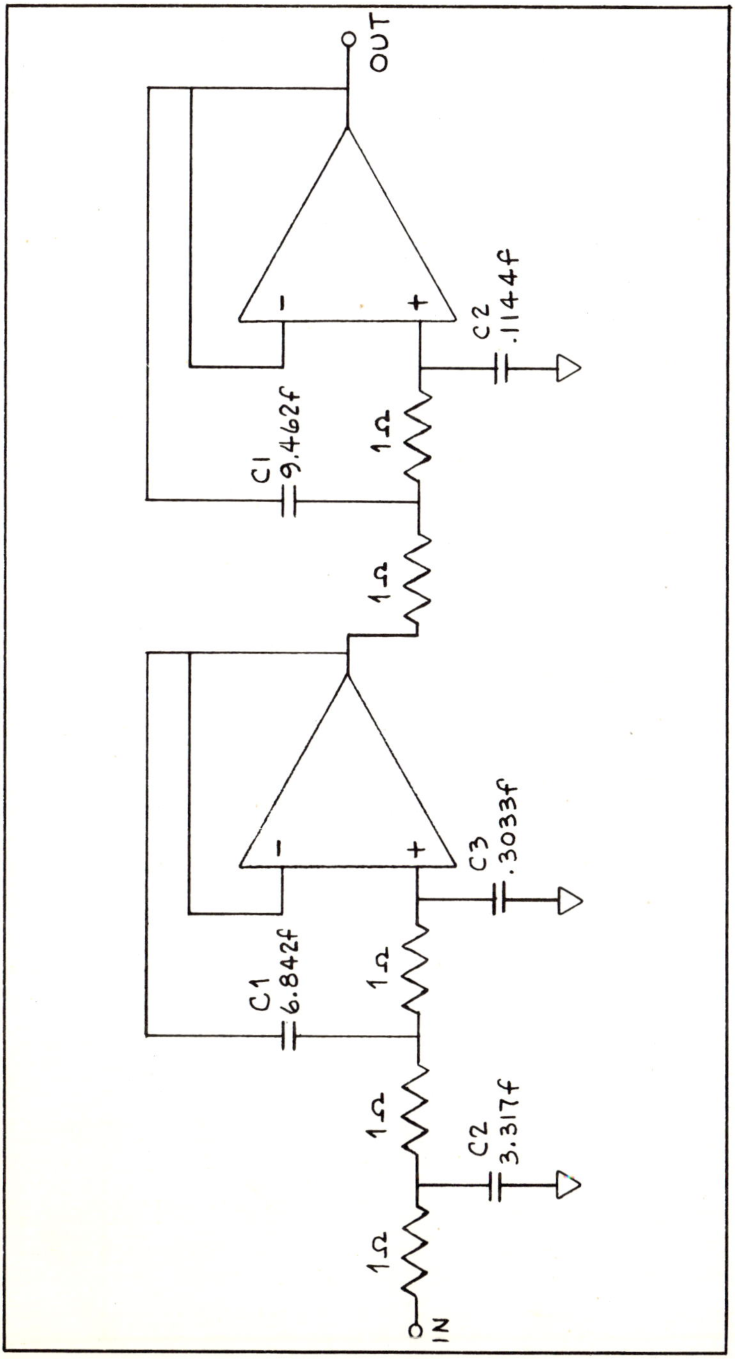

Figure 3-2a
Normalized N=5 .5db Chebyshev
Low-Pass Filter

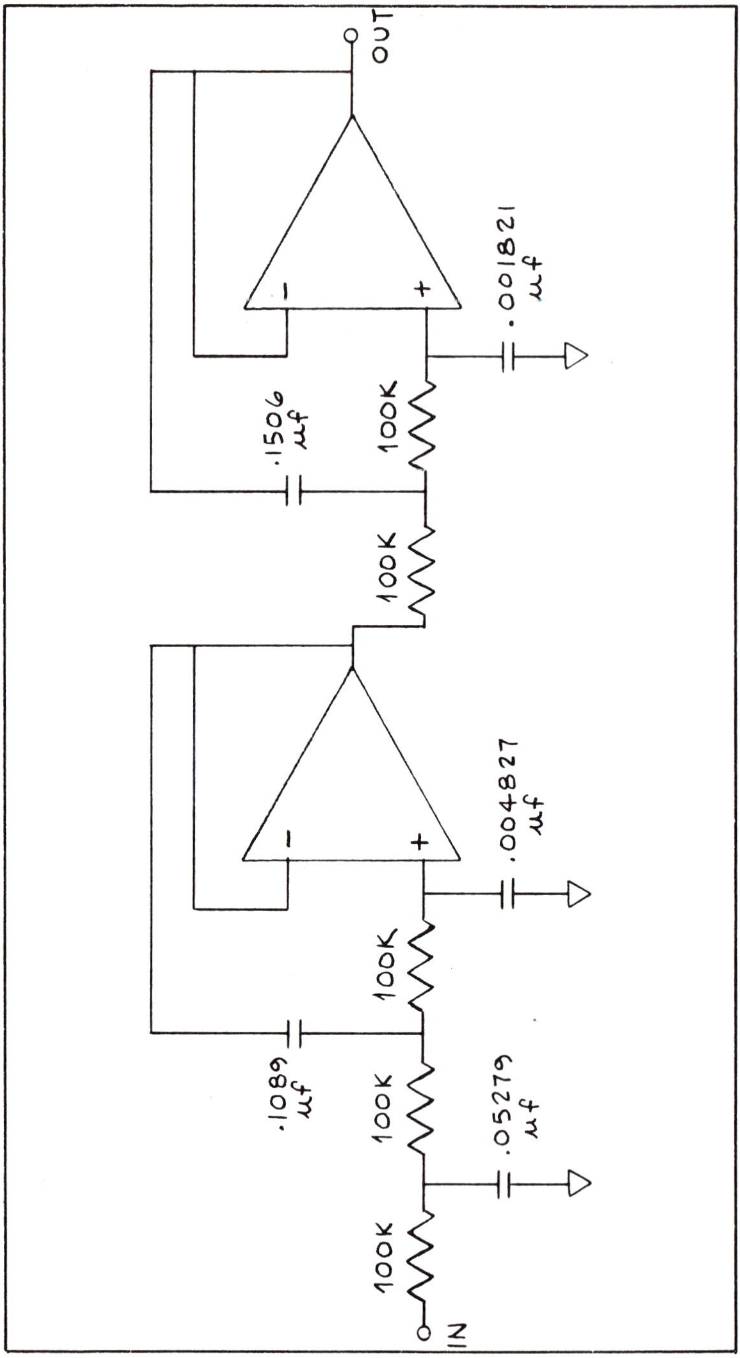

Figure 3-2b
Low-Pass Filter Scaled to
100Hz and 100K Ohm

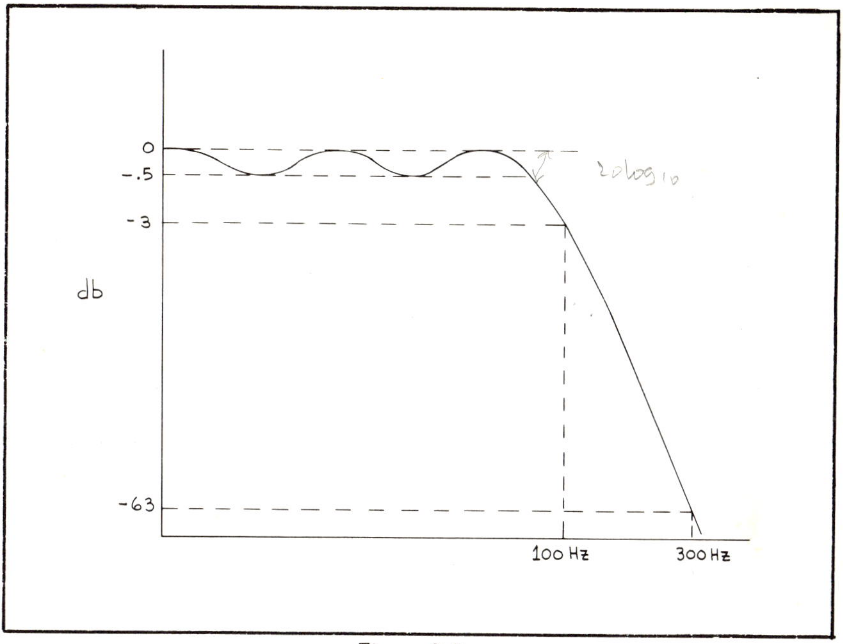

Figure 3-2c
Response of Example 3-1

The resistors are multiplied by Z resulting in equal resistors throughout of Z ohms.

A different value of Z could be used for each section to force practical capacitor values but the FSF must be the same for all sections.

Each section has a gain of unity at DC so the overall DC gain is unity. Individual sections may have a sharply peaked response which provides some gain in the passband. However the composite response of all the sections in cascade results in the desired Low-Pass response.

Care should be taken that the first section is driven by a voltage source having a source impedance which is negligible compared to the first resistor of the section. The driving source should provide a DC return to ground of 30K or less if a blocking capacitor is present. Since the filter output impedance is very low the response is not affected by the load providing that the amplifier has sufficient driving capability.

The following example, 3-1, demonstrates the design procedure of an All-Pole Low-Pass filter. Figure 3.2 presents the resulting filter.

Example 3-1

Given: Low-Pass Filter Requirement
3db at 100Hz
60db minimum at 300Hz

Result: Calculate Steepness Factor

$$A_s = \frac{F_s}{F_c} = \frac{300}{100} = 3 \qquad (2\text{-}5)$$

Select Normalized Low-Pass Filter
N = 5 .5db Chebyshev
Meets 60db minimum at 3 Radians

Frequency and Impedance Scale
Divide Capacitors by Z x FSF
Multiply Resistors by Z
where: $FSF = 2\pi F_c = 2\pi \times 100$
Z chosen at 100K

b. High-Pass Filter Design

High-Pass Filter design follows a similar procedure as the design of Low-Pass filters. However, after selection of a filter type, the normalized Low-Pass filter must be transformed into a normalized high—pass filter. This is accomplished by replacing each resistor, R, by a capacitor of 1/R farads. Each capacitor C, is replaced by a resistor of 1/C ohms.

The normalized high-pass filter is then frequency scaled to the desired cutoff F_c and simultaneously impedance scaled. As in the Low-Pass case the capacitors are divided by Z x FSF and the resistors are multiplied by Z. Z can be different for each section but the FSF must be uniform throughout the filter.

The filter gain is unity at the higher frequencies in the flat portion of the pass-band. The driving source impedance should be much less than the reactance of the first capacitor of the first filter section at the highest pass-band frequency of interest.

Example 3-2 and figure 3-3 illustrates the design of a high-pass filter.

Example 3-2

Given: High-Pass Filter Requirement
3db at 100Hz
60db minimum at 33.3Hz

Result: Calculate Steepness Factor

$$A_s = \frac{F_c}{F_s} = \frac{100}{33.3} = 3 \qquad (2\text{-}8)$$

Select Normalized Low-Pass Filter

N = 5 .5db Chebyshev
Meets 60db minimum at 3 Radians

Transform to Normalized High-Pass Filter

Replace each capacitor C by resistor of 1/C ohms
Replace each resistor R by capacitor of 1/R farads.

Frequency and Impedance Scale

Divide Capacitors by Z x FSF
Multiply Resistors by Z
where: $FSF = 2\pi F_c = 2\pi \times 100$
Z chosen at 100K

3.2 ELLIPTIC FUNCTION LOWPASS AND HIGHPASS FILTERS

The normalized Low-pass frequency response of an Elliptic function filter is shown in Figure 2-20 where R_{db} is the pass-band ripple in DB, A_{db} is the minimum stop-band attenuation in DB and ω_s is the Lowest stop-band frequency at which A_{db} occurs. All frequencies are in Radians per Second. The cutoff frequency is 1 Radian where the attenuation is R_{db} rather than 3db as in the case of the All-Pole Low-Pass filters.

The circuit schematic of an Elliptic-Function Low-Pass filter section is illustrated in Figure 3-4a. The total number of sections required for a complete filter is (N-1)/2 where N defines the order of the filter. Only odd order filters are tabulated in appendix 7.2 in order to provide the most efficient utilization of a fixed number of amplifiers.

An odd order Low-Pass transfer function contains only one real pole. This is realized by R5 and C5 which appear on the output section only.

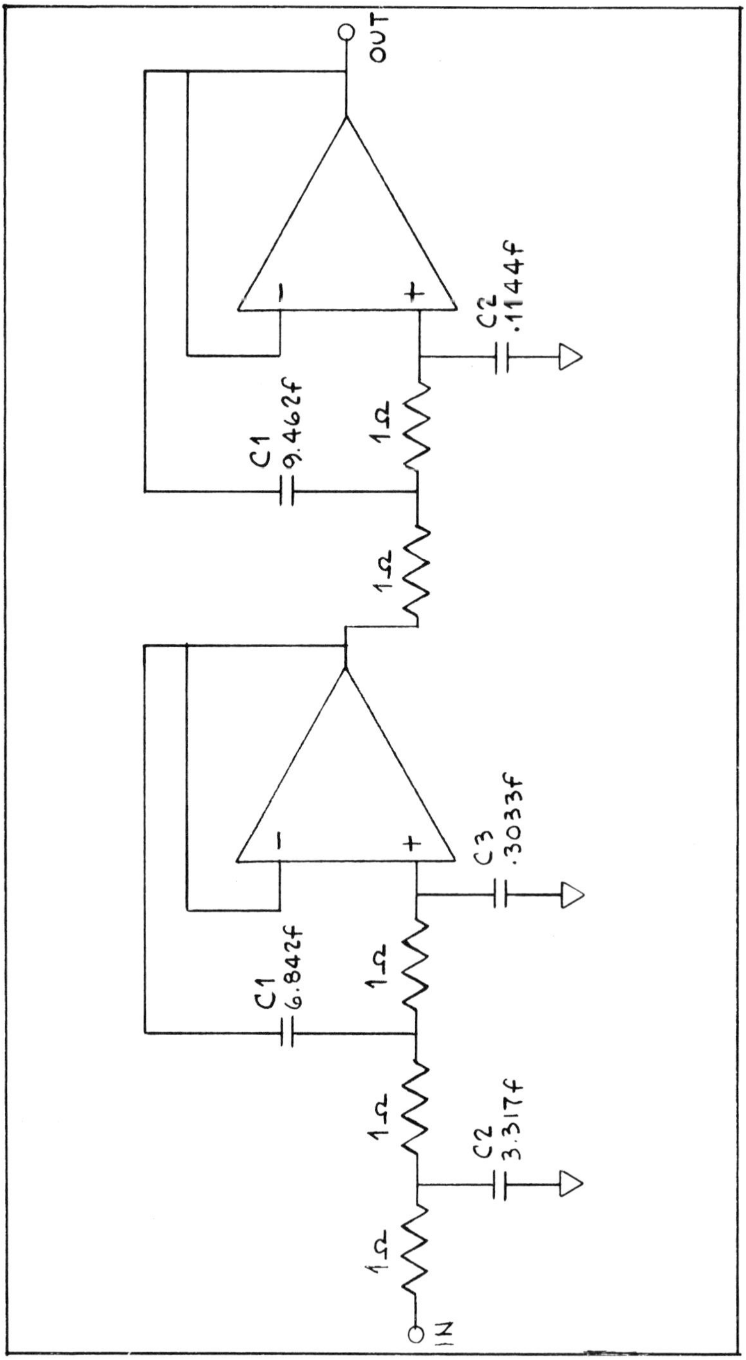

Figure 3-3a
Normalized N=5 .5db Chebyshev
Low-Pass Filter

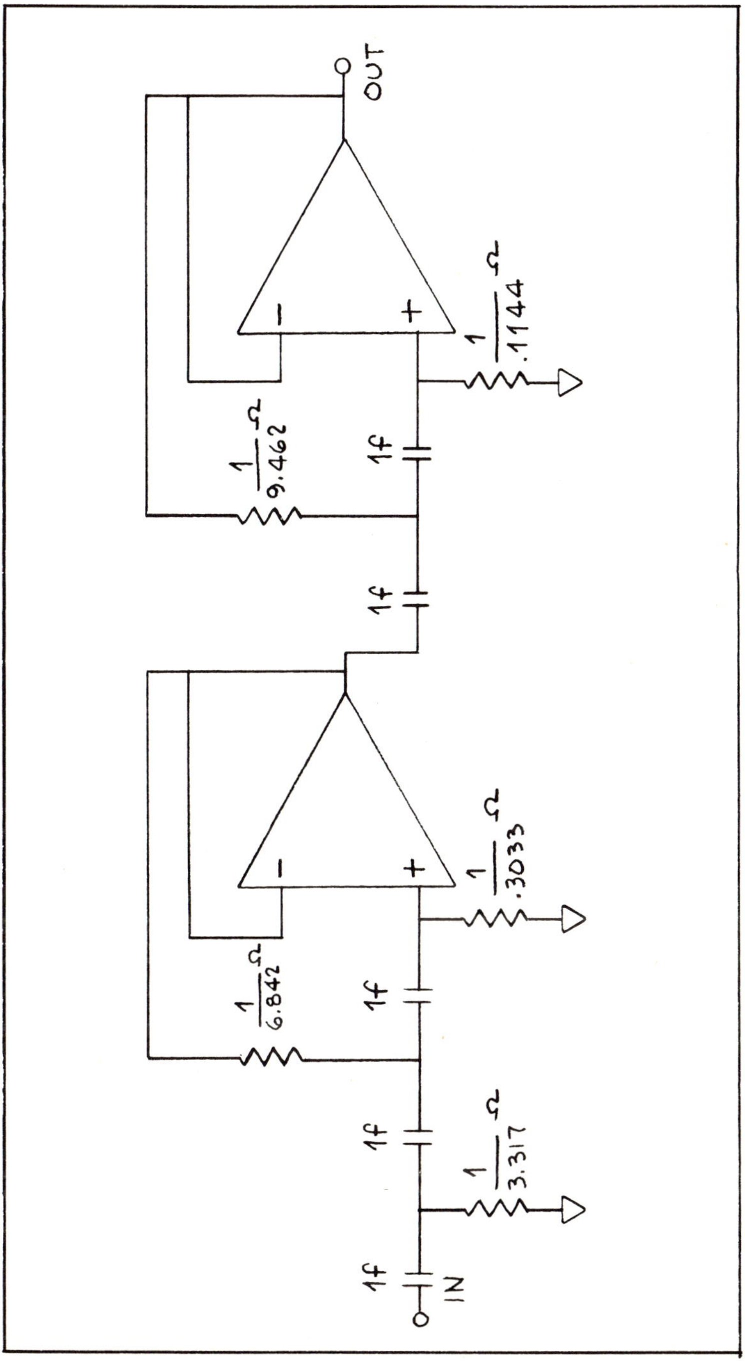

Figure 3-3b
Transformed Normalized High-Pass Filter

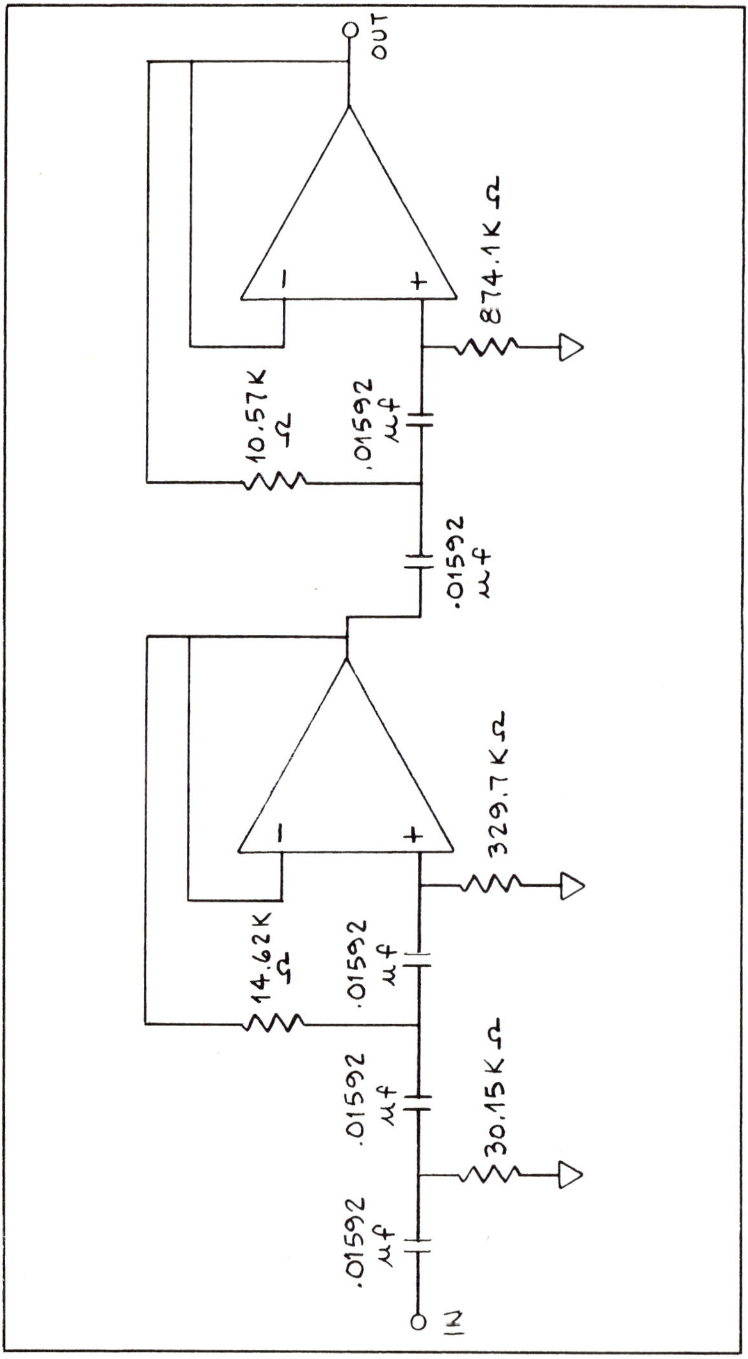

Figure 3-3c
High-Pass Filter Scaled to
100Hz and 100K Ohm

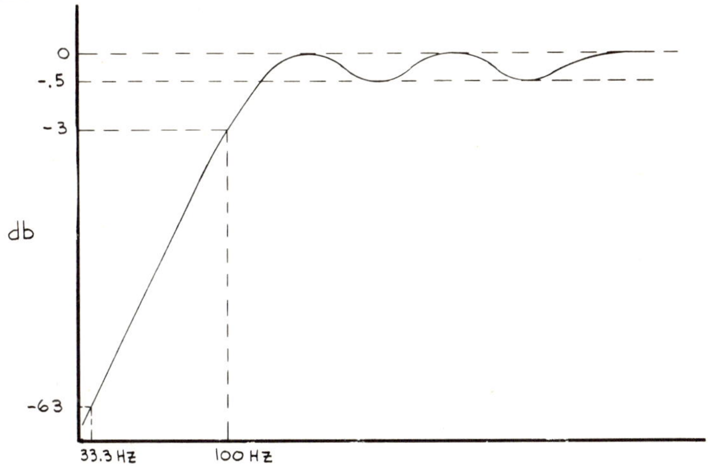

Figure 3-3d
Response of Example 3-2

Each section realizes a pair of complex conjugate poles and a pair of conjugate imaginary zeros. The co-ordinates are provided alongside the component values in appendix 7.2 where:

A = Real part of pole position.
B = Imaginary part of pole position.
Zero = Position of Zero

For the real pole section, R5 and C5, no imaginary part exists so only A is given.

Each section has a gain at DC. This value is also provided alongside the section element values. To compute the overall filter gain at DC, determine the product of all the individual section gains.
Within the feedback loop of each section a controlled amplification K is required from the non-inverting (+) amplifier input to the output. This is determined by resistors R6 and R7 having normalized values of 1 ohm and K-1 ohms respectively. In the event that K is less than 1, R6 is not used and R7 is a direct connection thus configuring the amplifier for a gain of unity. The attenuation K is achieved by splitting R4 into two resistors, R4a and R4b where:

$$R4a = (1\text{-}K)\ R4 \tag{3-1}$$

$$R4b = K x R4 \tag{3-2}$$

Figure 3-4b shows the modified configuration.

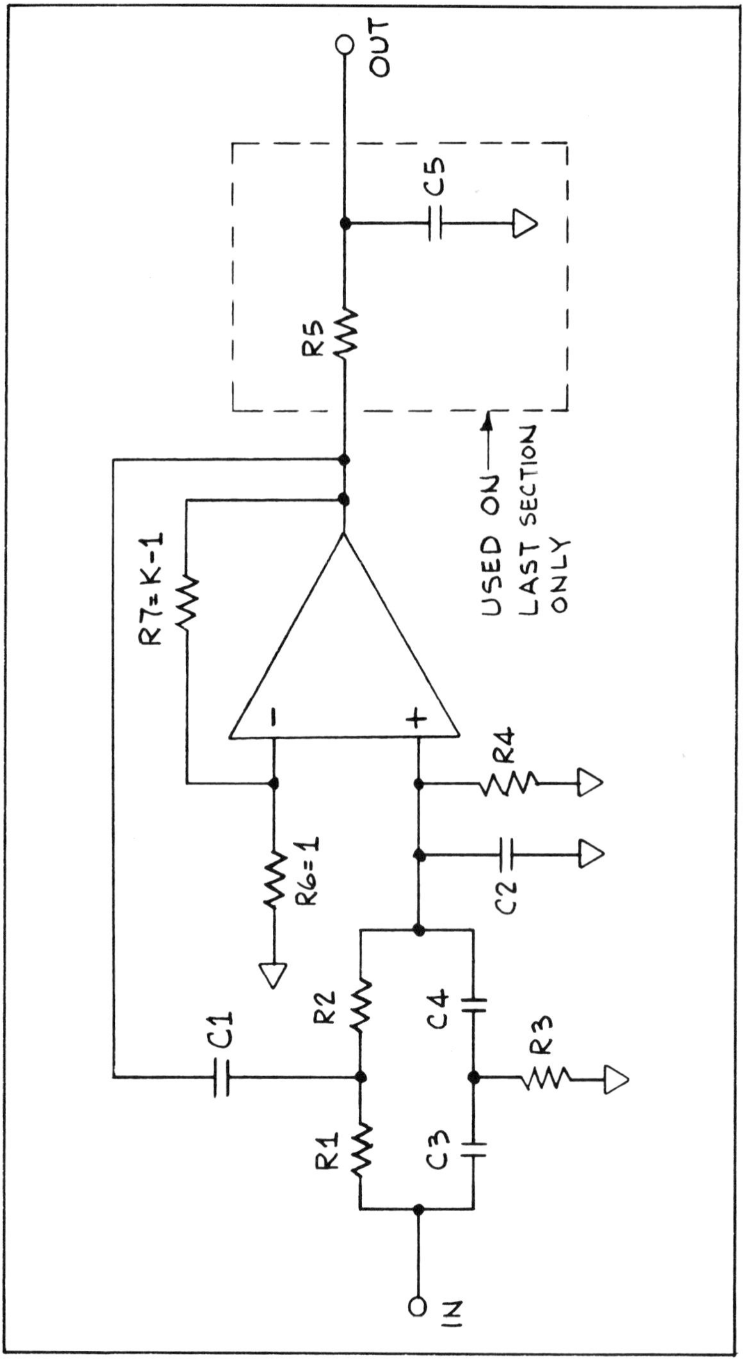

Figure 3-4a
Elliptic Function Normalized
Low-Pass Filter Section $K > 1$

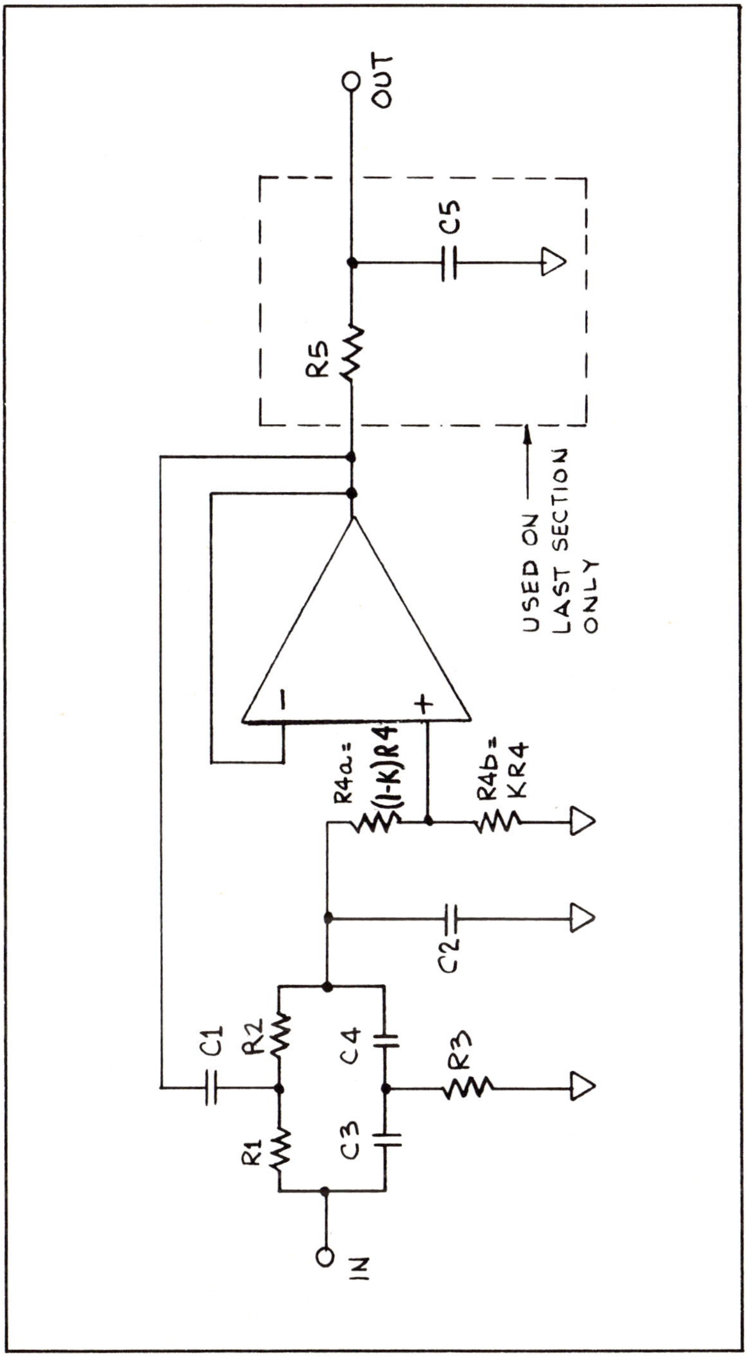

Figure 3-4b
Elliptic Function Normalized
Low-Pass Filter Section K < 1

a. Low Pass Filter Design

A given Low-Pass filter requirement must first be converted to a steepness Factor A_s. A Normalized Low-Pass Filter is then selected from section 7.2 that achieves the required passband to stop-band Transition by having an ω_s not more than A_s. The passband ripple R_{db} and minimum attenuation A_{db} should also satisfy the specification.

The selected filter is then scaled to the desired cutoff frequency F_c by dividing the capacitor values by Z x FSF. The resistors are multiplied by Z except for R6 and R7 which can be multiplied by any arbitrary factor since only the ratio of the two enters the design.

Each section can be denormalized by a different value of Z if desired providing that the FSF is constant throughout.

Similarly to the All-Pole Filters, the input section should be driven by a voltage source having a source impedance which is negligible compared to the input resistor, R1. Unlike the All-Pole filters the load impedance should be high in comparison to R5.

The frequency of infinite attenuation of each section is calculated by multiplying the tabulated zero by F_c.

The Low-Pass design procedure is shown in example 3-3 resulting in the filter of Figure 3-5.

Example 3-3

Given: Low-Pass Filter Requirement
.5db Maximum Ripple Below 1000Hz
35db Minimum Rejection Above 3000Hz

Result: Compute $A_s = \frac{3000}{1000} = 3$

Select Filter from Table 7.2

$N = 3$
$R_{db} = .28\text{db}$
$\omega_s = 2.924$
$A_{db} = 39.48\text{db}$

Frequency and Impedance Scale

Divide Cs' by Z x FSF
Multiply Rs' by Z
where: $FSF = 2\pi F_c$
$= 2\pi \times 1000$
Z chosen at 10K

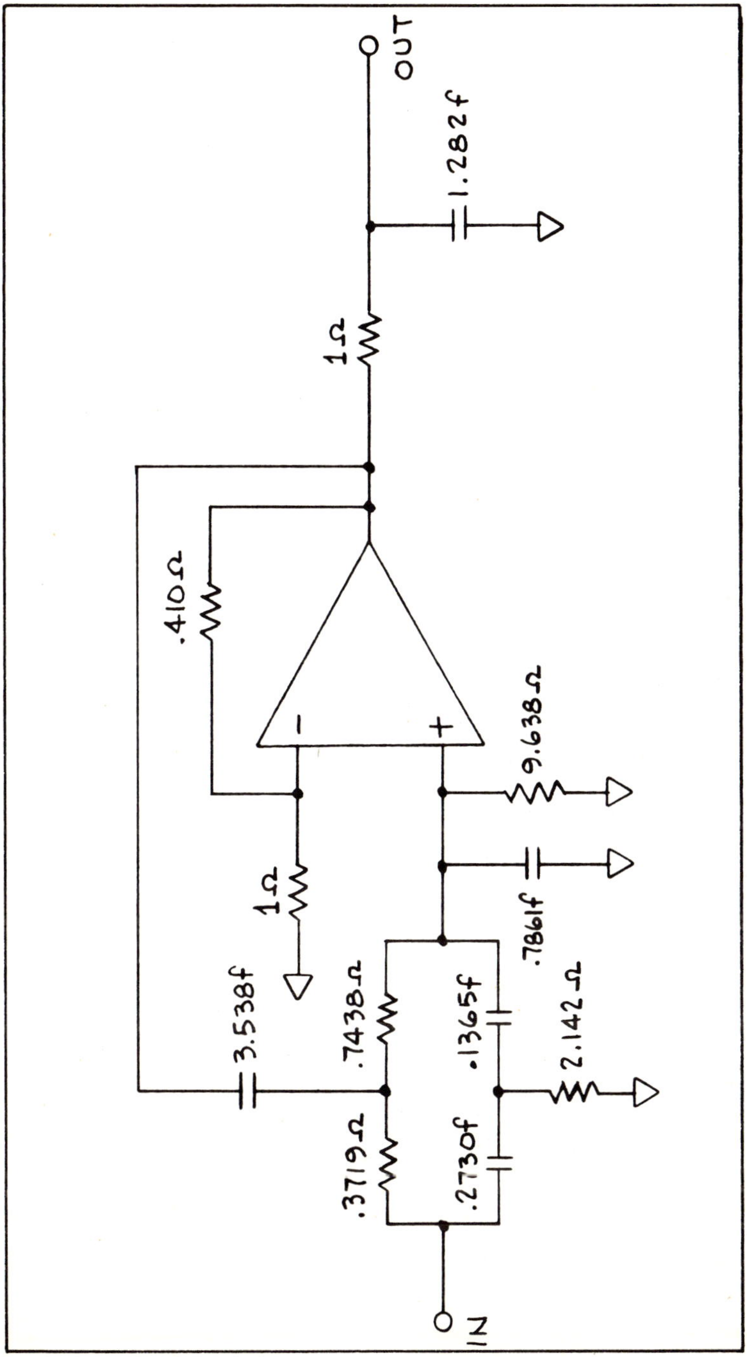

Figure 3-5a
Normalized Low-Pass Filter of Example 3-3

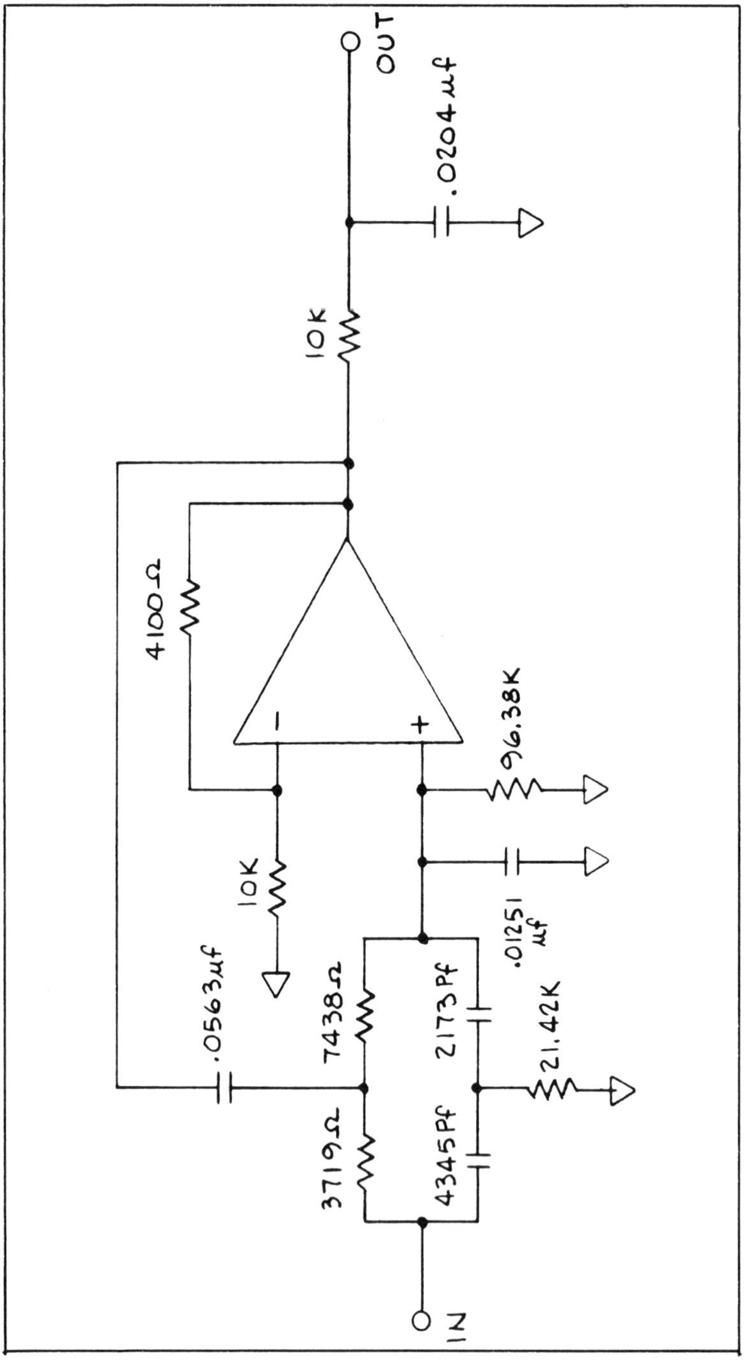

Figure 3-5b
Denormalized Low-Pass Filter
for Z = 10K and F_c = 1000Hz

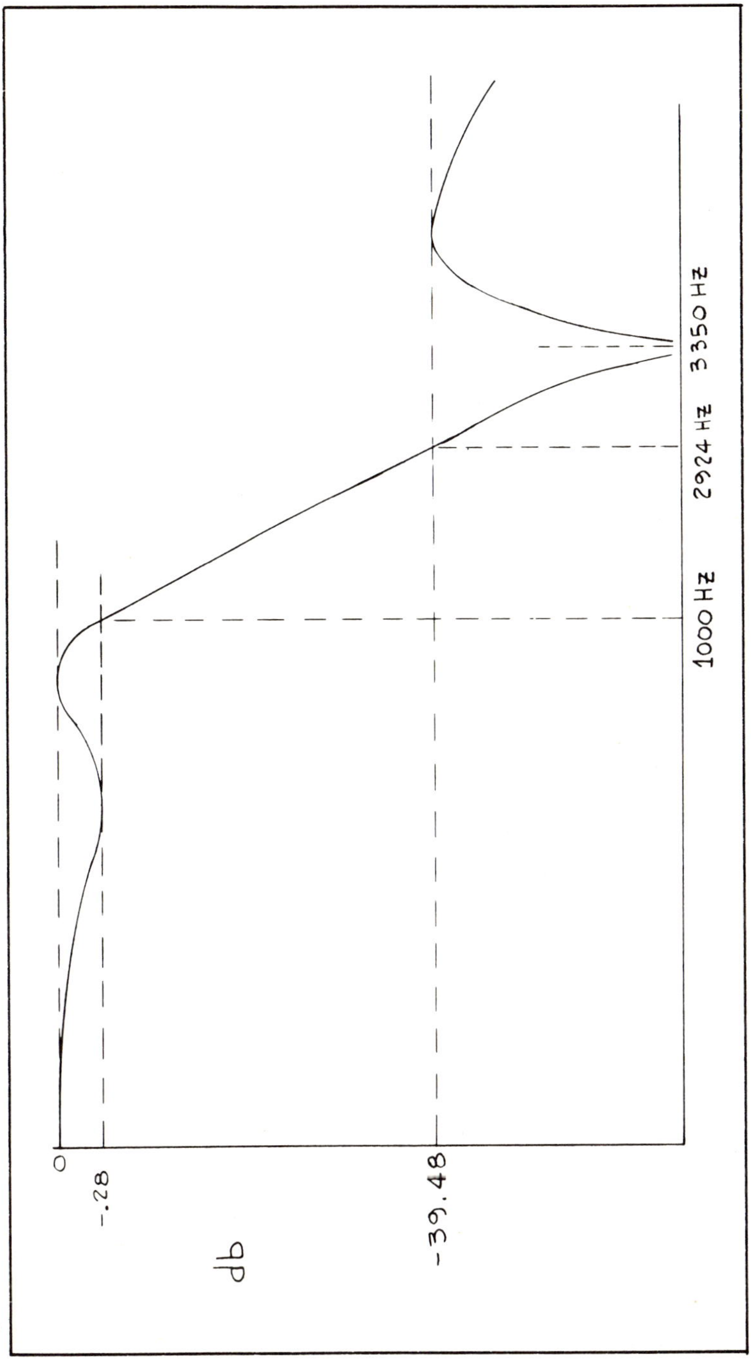

Figure 3-5c
Frequency Response of Example 3-3

The resulting filters frequency response is .28db at 1000Hz and 39.48db at 2924Hz.

b. High Pass Filter Design

High-Pass filters can be designed using the normalized Elliptic Function Low-Pass filter tables in appendix 7.2.

A_s must first be computed where for the High-Pass case:

$$A_s = \frac{F_c}{F_s} \qquad (2\text{-}8)$$

The tables are then examined for a filter having an ω_s not exceeding A_s and an R_{db} and A_{db} in accordance with the requirements. The selected normalized Low-Pass filter is converted to a normalized High-Pass filter by replacing each resistor, R, by a capacitor of 1/R farads (except R6 & R7) and replacing each capacitor C, by a resistor of 1/C ohms.

Figure 3-6a shows the Normalized High-Pass configuration for K greater than 1. For a K of less than 1 the configuration of Figure 3-6b is used.

The normalized High-Pass filter is frequency and Impedance scaled by dividing all capacitors by Z x FSF and multiplying all resistors by Z.

The filter gain is calculated identically to the Low-Pass case and is the product of the tabulated section gains. The source impedance of the driving source should be insignificant compared to the reactance of the input capacitor C3, at the highest pass-band frequency of interest. Example 3-4 and Figure 3-7 show the design procedure.

Example 3-4

Given: High-Pass Filter Requirement
.5db Maximum Ripple above 300Hz
35db Minimum Rejection at 100Hz & below

Result: Calculate Steepness Factor

$$A_s = \frac{F_c}{F_s} = \frac{300}{100} = 3$$

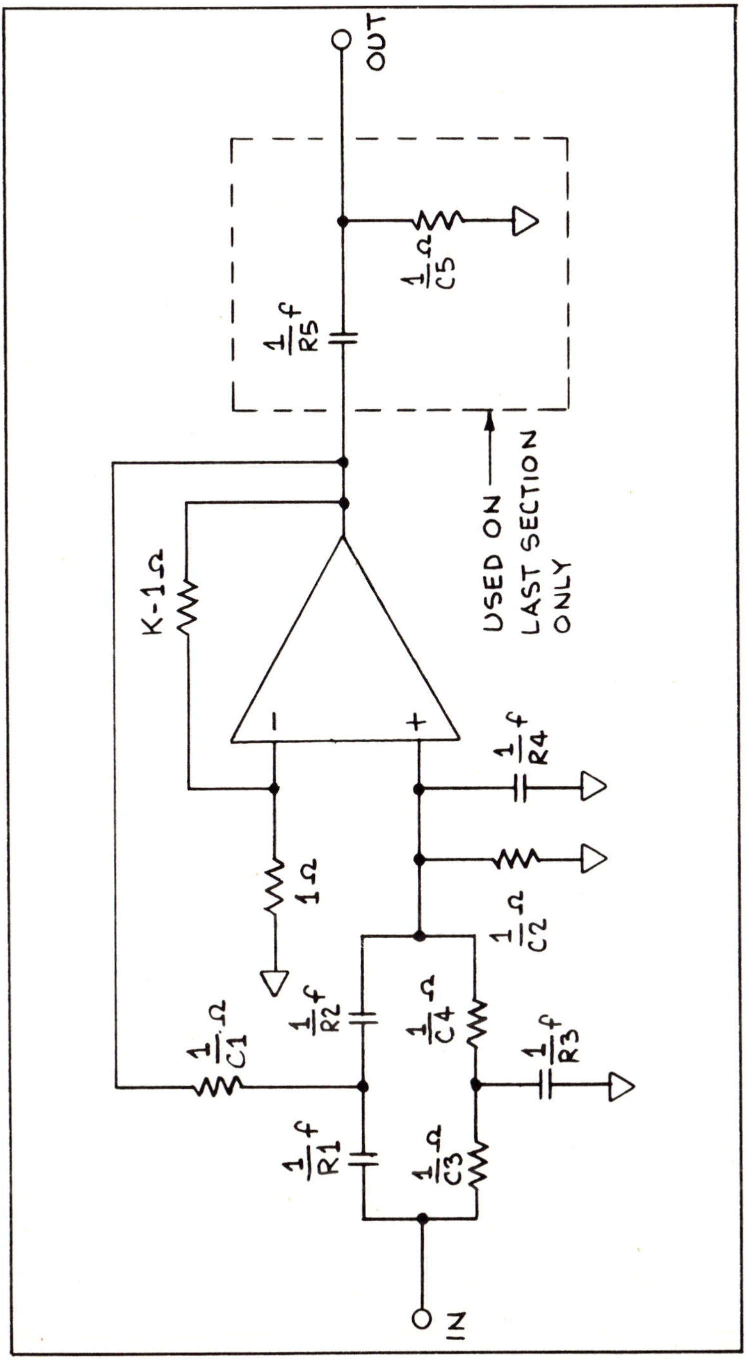

Figure 3-6a
Elliptic Function Normalized
High-Pass Filter Section K > 1

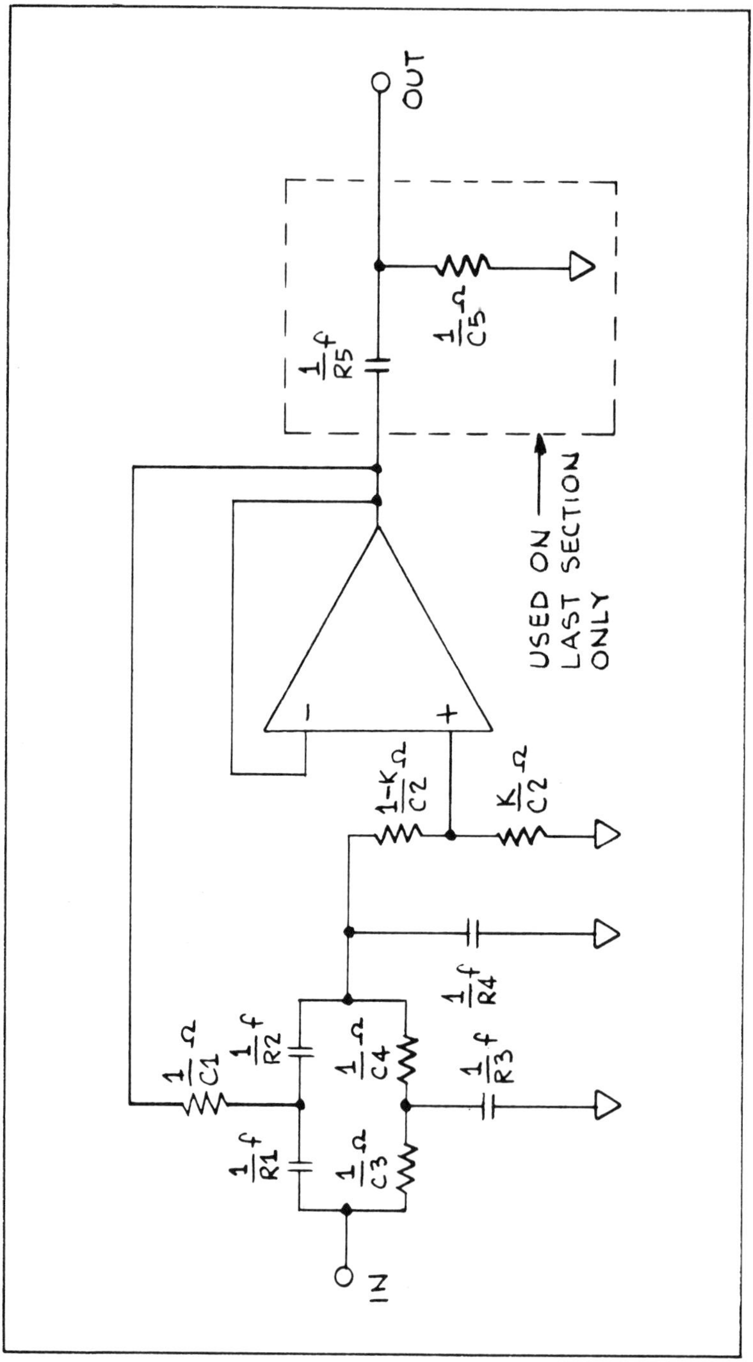

Figure 3-6b
Elliptic Function Normalized·
High-Pass Filter Section K < 1

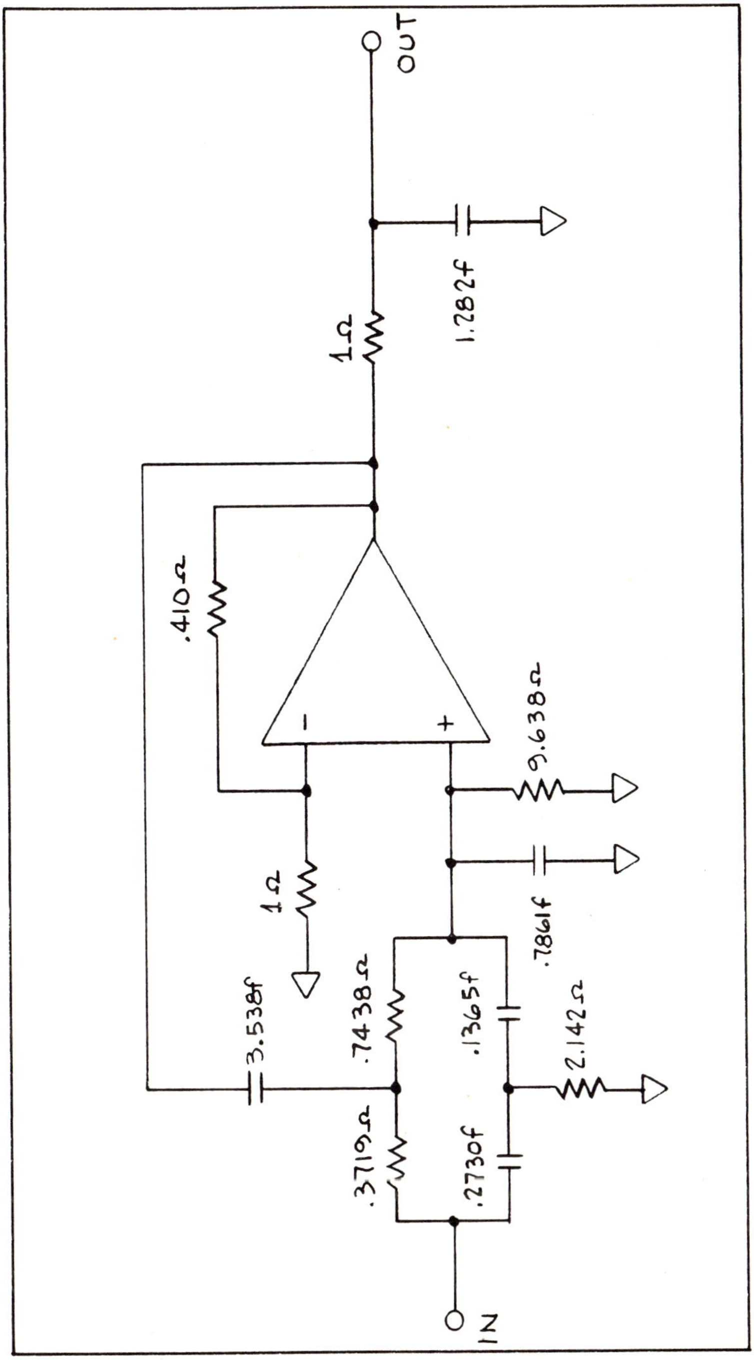

Figure 3-7a
Normalized Low-Pass Filter of Example 3-4

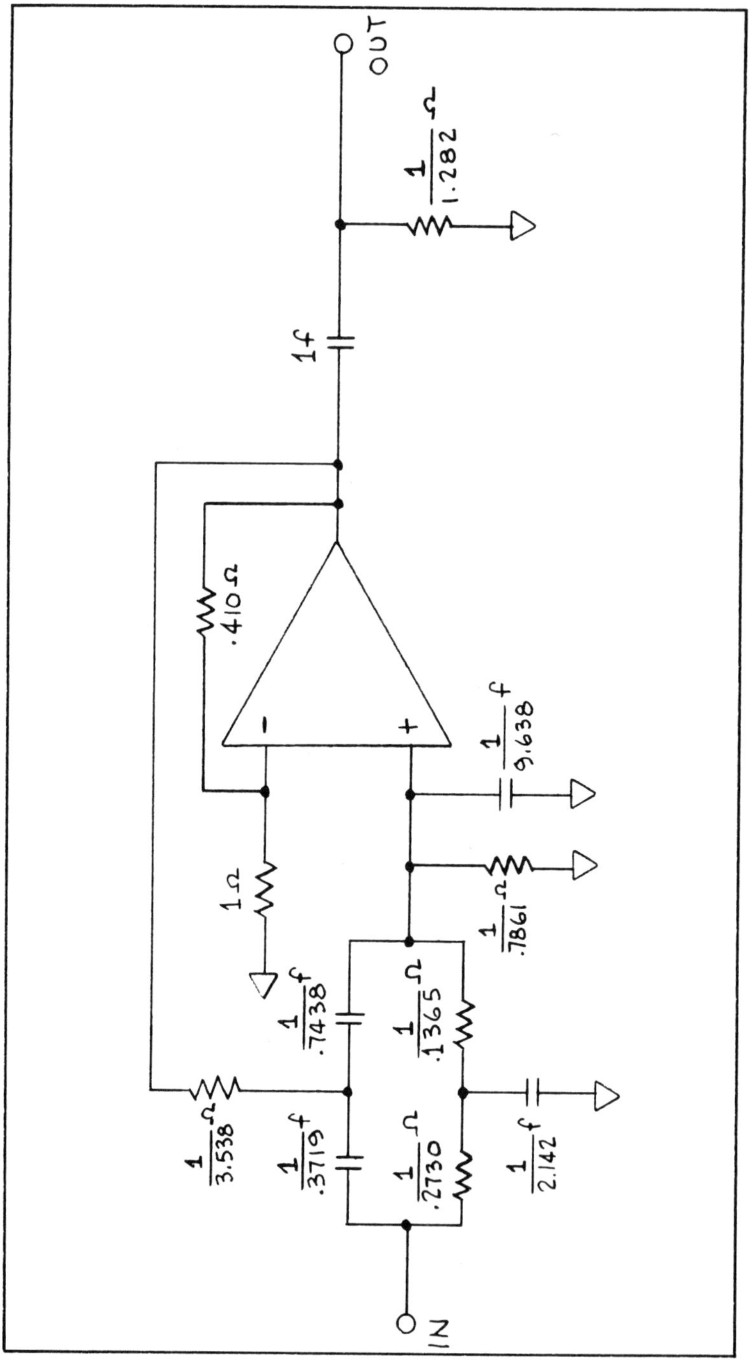

Figure 3-7b
Transformed Normalized High-Pass Filter

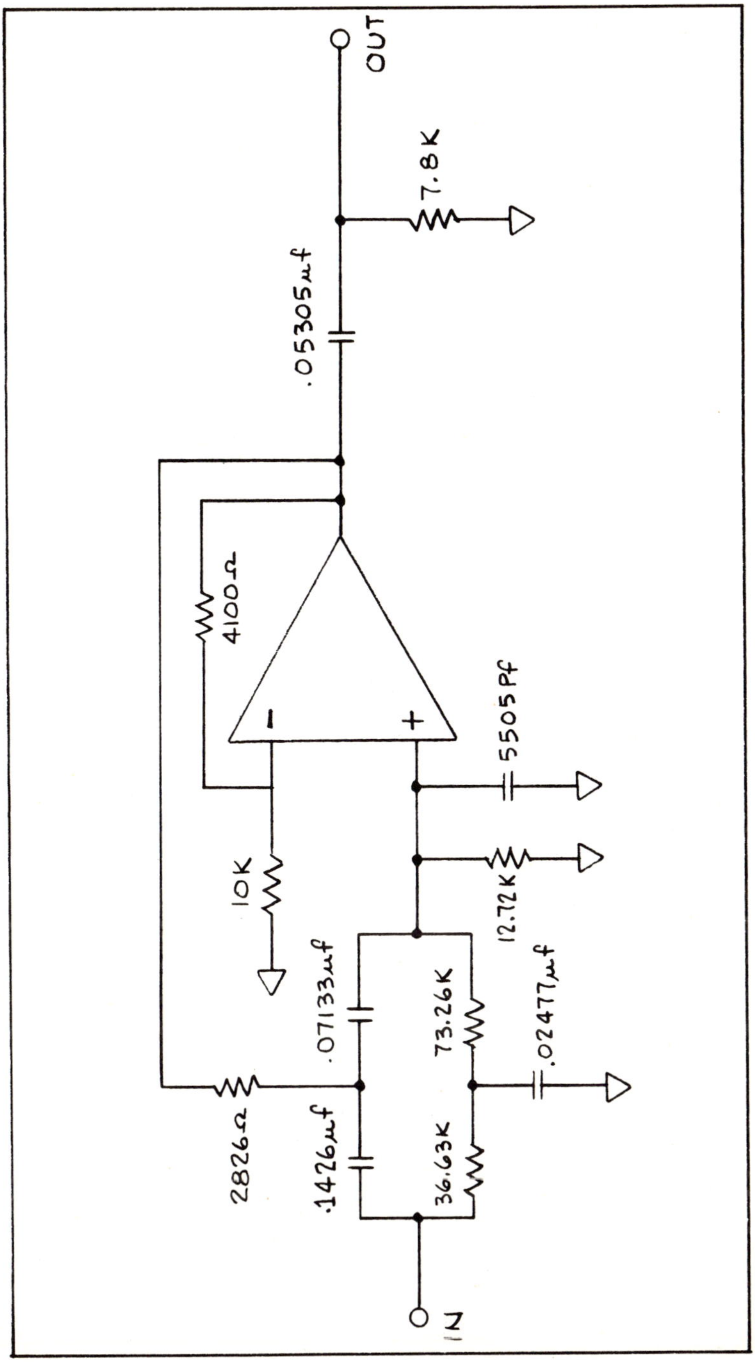

Figure 3-7c
Denormalized Filter for Z = 10K, F_c = 300Hz

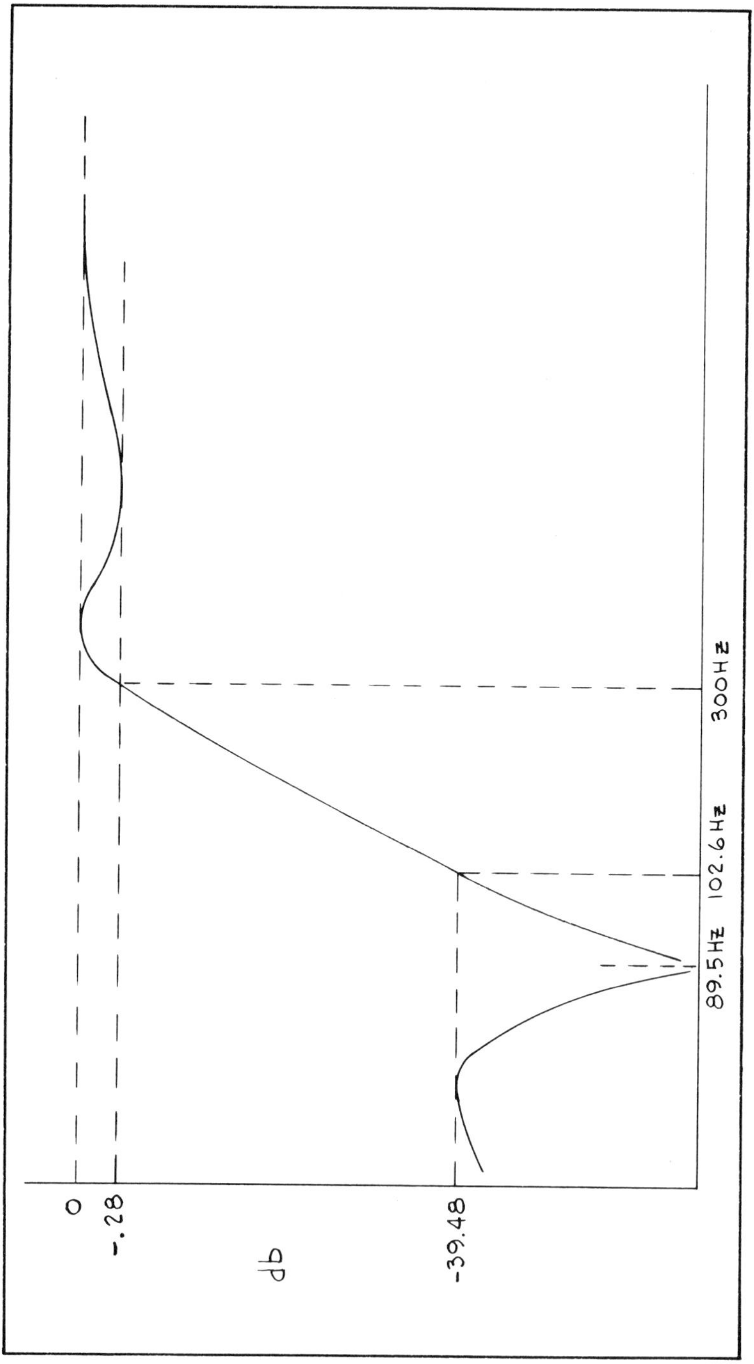

Figure 3-7d
Frequency Response of Example 3-4

Select Normalized Low-Pass Filter
from Table 7.2

$N = 3$
$R_{db} = .28db$
$\omega_s = 2.924$
$A_{db} = 39.48db$

Transform to Normalized High-Pass Filter

Replace each capacitor C by resistor of 1/C ohms
Replace each resistor R by capacitor of 1/R
farads except for R6 and R7

Frequency and Impedance Scale

Divide Capacitors by Z x FSF
Multiply Resistors by Z
where: $FSF = 2\pi\ F_c = 2\pi \times 300$
Z chosen at 10K

Expected Response

Since $\omega_s = 2.924$
Scaling the cutoff to 300Hz for .28db
results in 39.48db at 300/2.924
or 102.6Hz

In the process of making the Normalized Low-Pass to High-Pass transformation, the Zeros are transformed to reciprocal frequencies so that a Zero at ω_1 is transformed to $1/\omega_1$. The denormalized Zero occurs at F_c/ω_1.

3.3 WIDE-BAND BAND-PASS FILTERS

If the ratio of the upper cutoff frequency to the lower cutoff frequency is in excess of 1.5 the filter requirement can be considered Wide-Band. The specification is segregated into a Low-Pass filter requirement and a High-Pass filter requirement as in example 2-4 so that a cascade of Low-Pass and High-Pass filters will provide the required overall response.

If Elliptic Function filters are cascaded, an intermediate buffer stage is required to provide a high load impedance for the real-pole section of the first filter. Figure 3-8 illustrates two alternate buffer configurations which have a high input impedance. Figure 3-8a is a voltage Follower having unity gain. Figure 3-8b is a non-inverting amplifier which provides a gain of 1+R2/R1. R1 and R2 can be any convenient magnitude since only the ratio is of significance. All-Pole Low-Pass filters and High-Pass Filters can be cascaded arbitrarily without buffering.

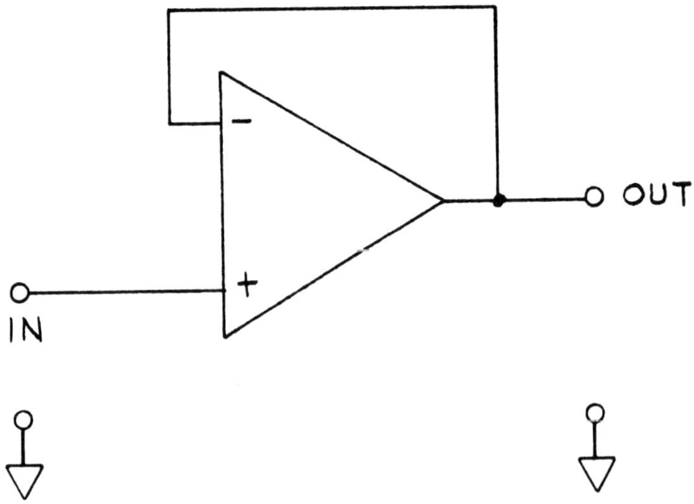

Figure 3-8a
Voltage Follower

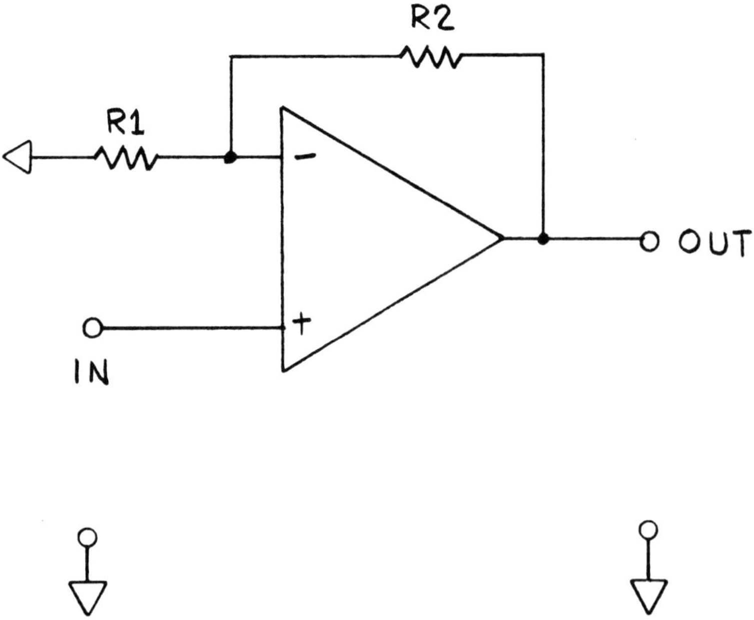

Figure 3-8b
Non-Inverting Amplifier

The criteria for defining a filter as wide-band i.e. an upper cutoff frequency to lower cutoff frequency ratio of 1.5, is somewhat arbitrary. An exact determination of the results of a wide-band approach can be made from the following reasoning.

The response of a cascade of two filters is equal to the algebraic sum of the individual frequency responses in DB. Therefore the loss in the pass-band can be estimated by combining the Low-Pass and High-Pass attenuation.

Low-Pass and High-Pass filter types that behave in a rectangular manner in the vicinity of cutoff can be combined for narrower requirements than filters having a rounded response. Figure 3-9 shows the results of attempting to generate too narrow a band-pass filter using a Low-Pass and High-Pass cascade approach. The additional loss at center frequency represents reduced relative stop-band attenuation.

Chapter 4 discusses the valid approach for narrow band-pass filters.

3.4 WIDE-BAND BAND-REJECT FILTERS

Band-Reject filters can be produced from Low-Pass and High-Pass filters by a parallel configuration. A graphical analysis similar to figure 3-9 is cumbersome since the parallel combination requires summation of the individual filter outputs including the phase angles. However, if the cutoffs are far enough apart one filter will be well in the stop-band at the pass-band frequencies of the other filter therefore minimizing any interaction after summation.

Typically if the ratio of upper cutoff frequency to lower cutoff frequency is greater than 1.5 a practical approach is the division of the requirement into separate Low-Pass and High-Pass specifications. The necessary filters are then directly paralled at the input and summed at the output.

Figure 3-10 illustrates two methods of combining output. The circuit of Figure 3-10a is intended for All-Pole filter types and has a gain of A where A and R may be any convenient value.

Elliptic Function Low-Pass and High-Pass filters can be combined using the configuration of figure 3-10b. The real-pole components of the last section of each filter are absorbed in this schematic and are therefore not necessary within the filter. R_A and C_A are the denormalized values of R_5 and C_5 of figure 3-4 for the Low-Pass

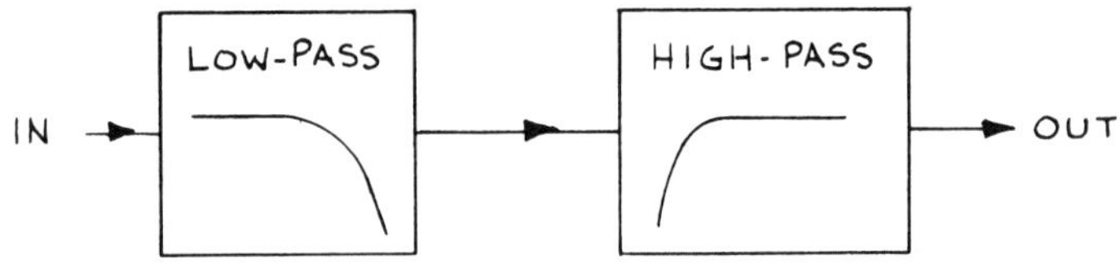

a) Low-Pass High-Pass Cascade

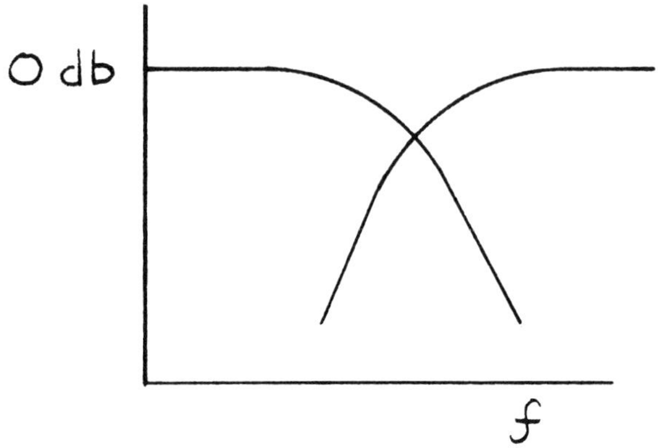

b) Composite Response

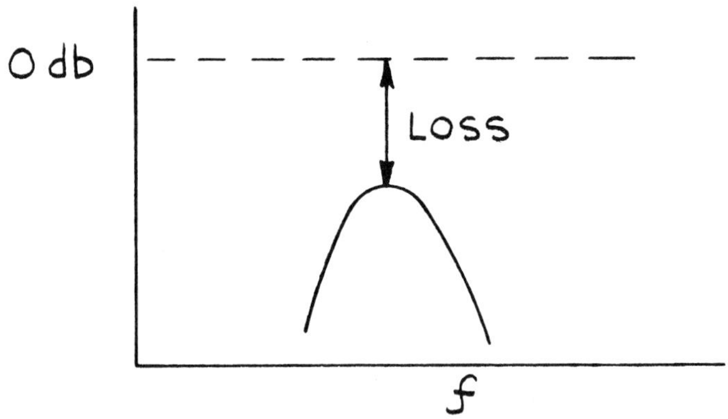

c) Algebraic Sum of Attenuation

Figure 3-9
Limitation of Wide-Band Approach
to Narrow-Band Filters

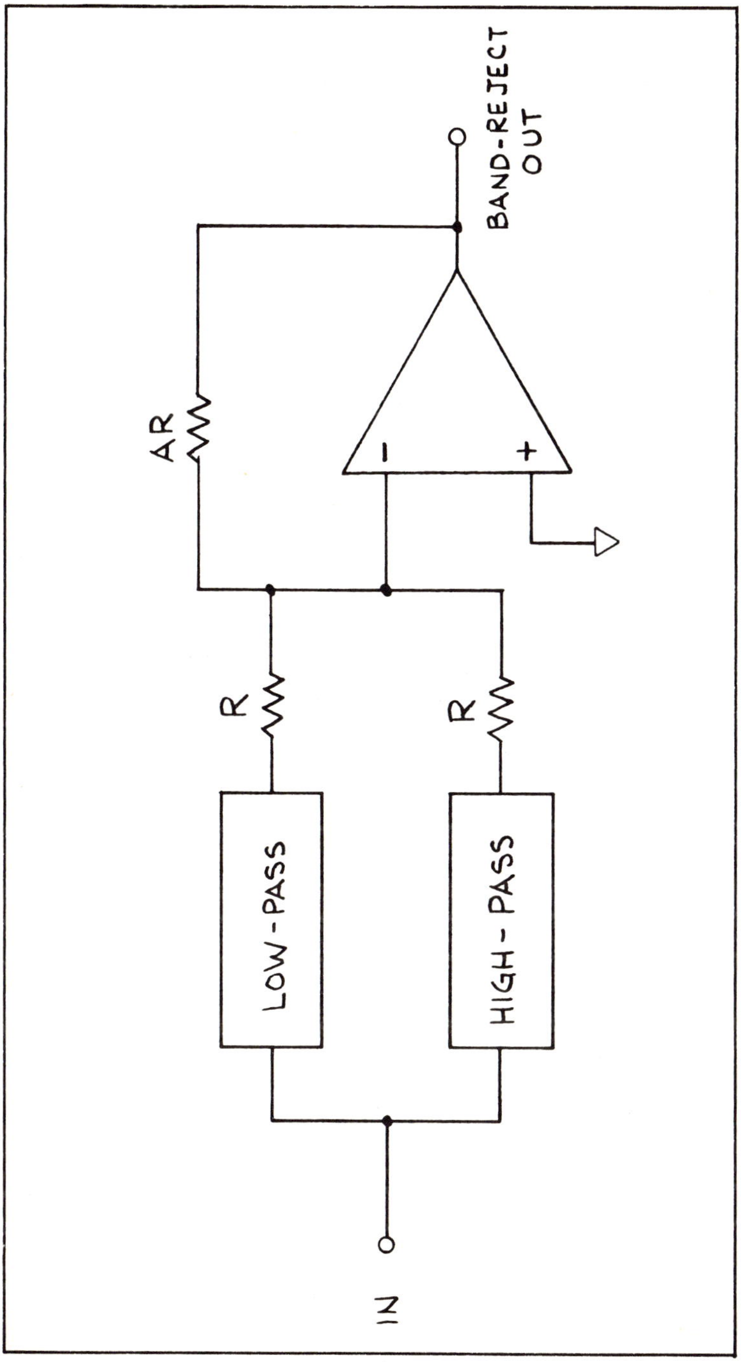

Figure 3-10a
Method for Combining All-Pole Filters

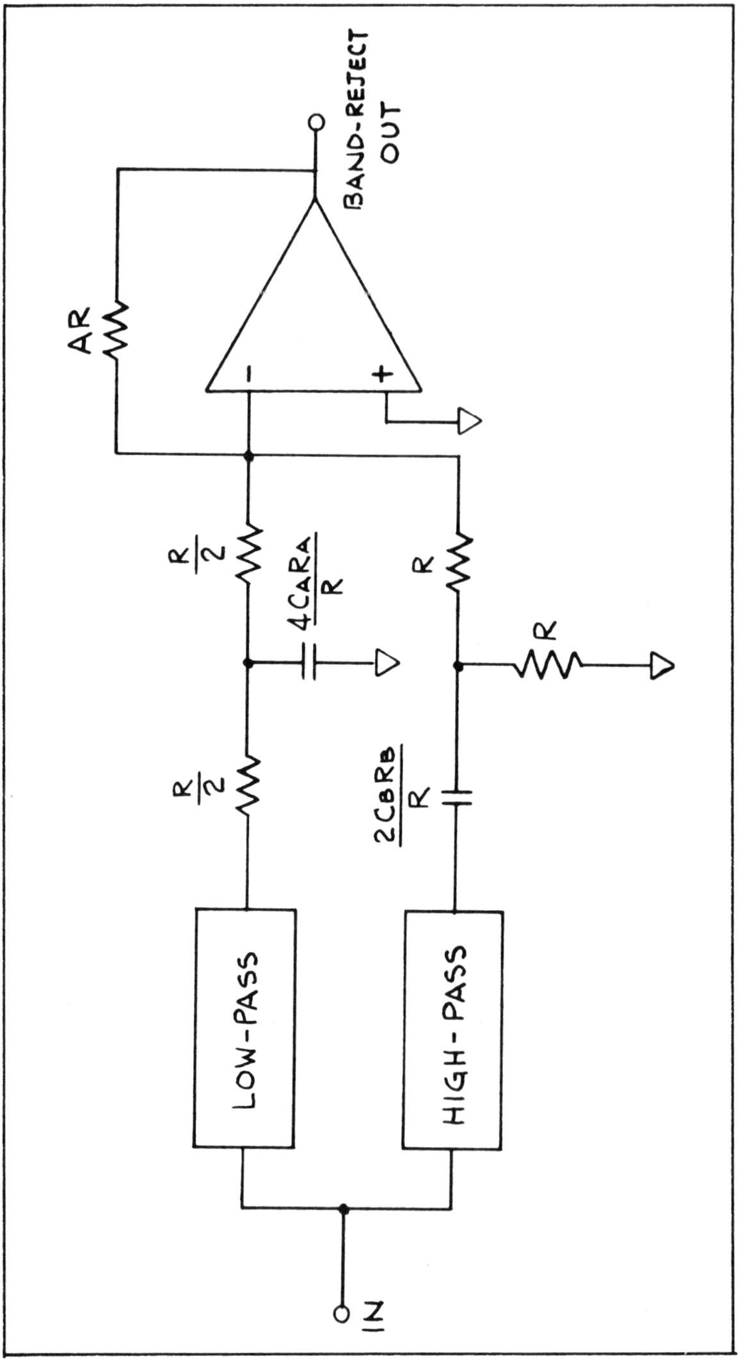

Figure 3-10b
Method for Combining
Elliptic-Function Filters

filter. The High-Pass filter denormalized real-pole values are R_B and C_B. The circuit provides a gain of A where A and R may be as desired.

Separation of a Band-Reject Filter specification into individual Low-Pass and High-Pass requirements is illustrated in the following example:

Example 3-5

Given: Band-Reject Filter Requirement

.5db Maximum Ripple below 100Hz and above 3000Hz
35db Minimum Rejection between 300Hz and 1000Hz

Result: Low-Pass Requirement

.5db Maximum Ripple below 100Hz
35db Minimum Rejection above 300Hz

High-Pass Requirement

.5db Maximum Ripple above 3000Hz
35db Minimum Rejection below 1000Hz

The reader is referred to section 5.1 for a discussion of Active Notch Networks for narrow-band requirements.

CHAPTER 4

Selective Band-Pass Filters

4-1 SINGLE POLE-PAIR REALIZATIONS

Narrow Band-Pass Filters are generated by transforming a Low-Pass transfer function into a Band-Pass transfer function and then realizing the Band-Pass Transfer Function with a cascade of active sections. As a result of this transformation each Low-Pass pole results in a pair of band-pass poles. ex: A 5 pole Low-Pass results in 5 Band-Pass pole-pairs.

A Band-Pass pole-pair provides a response equivalent to a single tuned circuit. It has been found convenient to define this pole-pair in terms of its two band-pass parameters, F_0' and Q where F_0' is the geometric center frequency in Hz and Q is the well known selectivity factor associated with tuned circuits which can be defined as:

$$Q = \frac{F_0'}{F_{bw}} \qquad (4\text{-}1)$$

where F_{bw} is the 3db bandwidth in Hz.

a. Single Amplifier Realization

Figure 4-1 illustrates a circuit for realizing a Band-Pass pole-pair. This configuration features a minimum number of components, simple tuning and low circuit sensitivity to component tolerances. However, amplifier open loop gain requirements are proportional to Q^2 therefore placing a restriction on realizable Qs's to below 20 typically. The subject is further discussed in section 6.2.

The overall circuit gain is $2Q^2$ at the section F_0'. It should be understood that this F_0' is determined by the co-ordinates of the particular

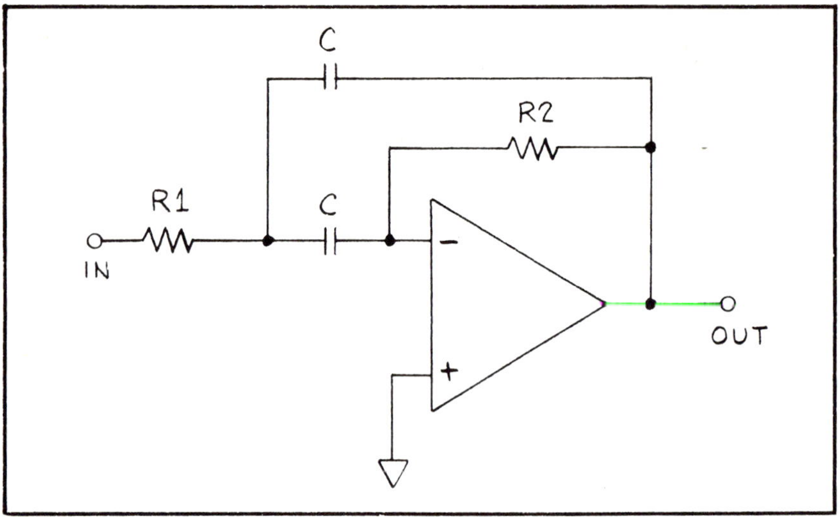

Figure 4-1
Single Operational Amplifier
Pole-Pair Realization

pole-pair and is not necessarily the center frequency of the composite filter. Therefore the section gain may be less than $2Q^2$ at the filter center frequency where the section may be off from resonance.

The section gain at the composite filter center frequency F_0 can be computed by:

$$A_0 = \frac{2Q^2}{\sqrt{1 + \left(\frac{2Q\Delta F}{F_0'}\right)^2}} \tag{4-2}$$

where ΔF is the frequency difference between section F_0' and filter F_0. When section frequency F_0' is equal to filter frequency F_0, ΔF is zero so equation 4-2 reduces to $A_o = 2Q^2$.

The element values of figure 4-1 are calculated from the band-pass parameters using the following relationships:

$$R_2 = \frac{Q}{\pi F_0' C} \tag{4-3}$$

$$R_1 = \frac{R_2}{4Q^2} \tag{4-4}$$

C is arbitrary and may be any convenient capacitor value.

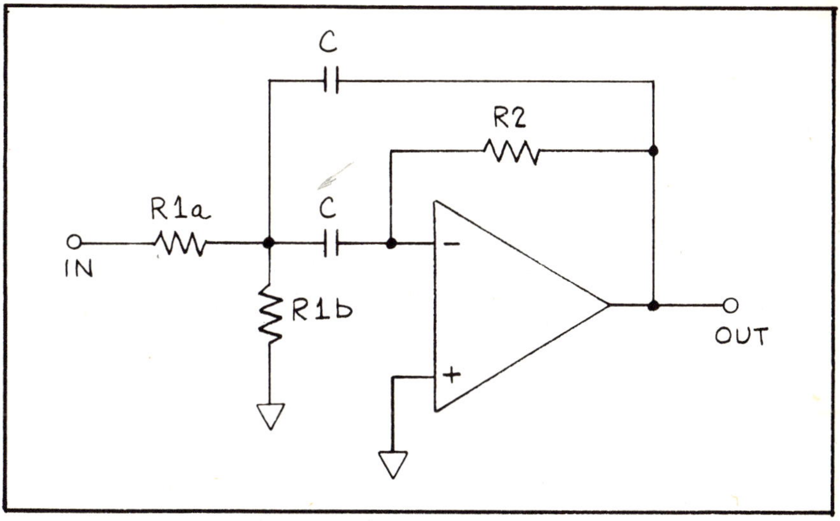

Figure 4-2
Reduced Gain Configuration

Equation 4-2 indicates that extremely high gains occur for moderate Qs' because of the Q^2 proportionality. This may not be desirable since amplifier clipping at the output may occur for high input levels. Figure 4-1 can be modified to provide a gain of A at Filter center frequency F_o. This is accomplished by providing a voltage divider at the input as illustrated in Figure 4-2. The parallel combination of R_{1a} and R_{1b} is equal to R_1. R_{1a} and R_{1b} can be calculated by:

$$R_{1a} = \frac{A_0}{4Q\pi F_0' CA} \tag{4-5}$$

$$R_{1b} = \frac{R_{1a}}{\frac{A_o}{A} - 1} \tag{4-6}$$

A can be any desired gain but must not exceed A_0.

b. State-Variable Pole-Pair Configuration

The high open-loop amplifier gain requirements of figures 4-1 and 4-2 become prohibitive for Qs' in excess of 20. A solution is to use the State-Variable realization which offers Low circuit sensitivity to component tolerances and less stringent amplifier gain requirements. This occurs at the expense of two more operational amplifiers

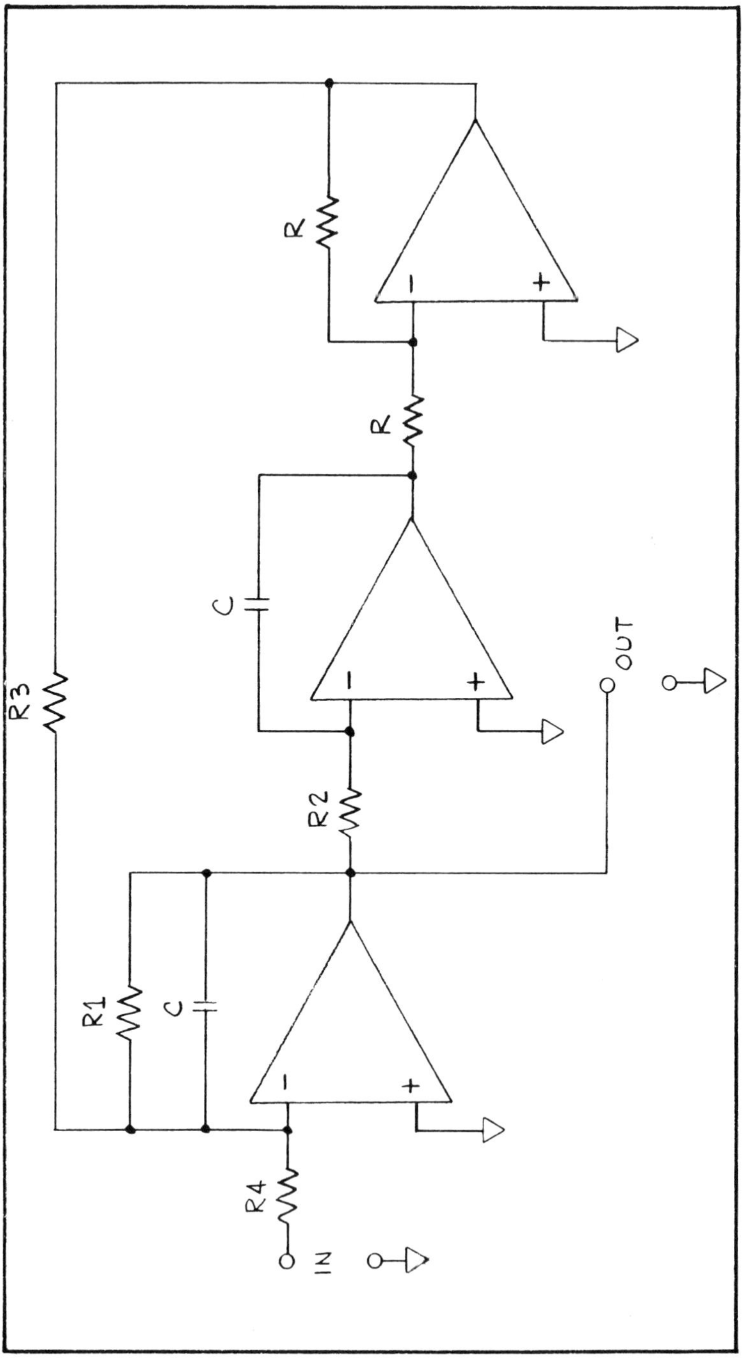

Figure 4-3
State-Variable Band-Pass Pole-Pair

and some additional components. However, the increased complexity is more than justified by the benefits for medium Q and high Q realizations.

Figure 4-3 illustrates the State-Variable configuration for a band-pass pole-pair. The component values are computed as follows:

$$R_1 = \frac{Q}{2\pi F_0' C} \tag{4-7}$$

$$R_2 = R_3 = \frac{R_1}{Q} \tag{4-8}$$

$$R_4 = \frac{R_1}{A\sqrt{1+\left(\frac{2Q\Delta F}{F_0'}\right)^2}} \tag{4-9}$$

When $F_0' = F_0$, equation 4-9 reduces to $R_4 = R_1/A$.

R and C of figure 4-3 may be arbitrarily selected. The desired section gain at composite filter center frequency F_0 is A. At section center frequency F_0' the gain is R_1/R_4.

This configuration offers unique adjustment capability where section resonant frequency, Q and gain can be independently adjusted. Adjustment techniques are discussed in section 4-3.

The decision to use a single operational amplifier configuration or the State-Variable circuit may not be clearcut since Q may not be the only criteria. An additional factor is the frequency range of operation. Sometimes Q's below 20 may require implementation by the State-Variable technique if the frequencies occur in the upper range of the operational amplifier where gain rolloff and amplifier phase-shift may be detrimental. For a discussion see section 6-2.

A particular band-pass filter contains N pole-pairs where the spread of pole Qs' may warrant use of both single operational amplifier and State-Variable configurations in cascade.

4.2 AN ALL-POLE BAND-PASS FILTER DESIGN PROCEDURE

Design of active band-pass filters is achieved by following these steps:

a) Convert the specification to a geometrically symmetrical band-pass filter requirement as described in section 2.2c.

b) Calculate A_s, the steepness factor, using equation 2-14, and select a Normalized Low-Pass filter type from Chapter 2.

c) Transform the normalized Low-Pass poles tabulated in appendix table 7-3 into band-pass pole-pairs.

d) Select a band-pass configuration from section 4-1 and compute the component values.

The band-pass pole-pair sections are then cascaded and may require tuning to acheive a more precise response.

Table 7.3 provides a tabulation of the roots of the polynomials representing the transfer-functions outlined in chapter 2. These roots or poles occur in two basic forms, real and complex. A real pole occurs singly and is therefore transformed into a single band-pass pole-pair. A complex pole occurs as a complex-conjugate pair which transforms into two band-pass pole-pairs. To perform a low-pass to band-pass pole transformation initially calculate

$$Q_c = \frac{F_0}{F_{bw}} \tag{4-10}$$

Where F_o is the filter geometric center frequency and F_{bw} is the filter 3db bandwidth.

a. Real-Pole Transformation

A normalized real pole having a real part of magnitude α is transformed into a complex band-pass pole-pair with the following band-pass parameters:

$$Q = \frac{Q_c}{\alpha} \tag{4-11}$$

$$F_o' = F_o \tag{4-12}$$

A real Low-Pass pole transforms into a band-pass pole-pair having a center frequency F_o' exactly equal to the composite filter center frequency F_o.

Therefore, ΔF is zero in equations 4-2 and 4-9 for computation of component values of the Band-Pass pole-pair section.

b. Complex Pole Transformation

A complex pole occurs as a complex-conjugate pair expressed in the form:

$$-\alpha \pm J\beta$$

where α is the real part and β is the imaginary part. Upon being given α, β, Q_c and F_o the following series of calculations lead to the two bandpass pole-pair Q and F_o' values.

$$C = \alpha^2 + \beta^2 \tag{4-13}$$

$$D = \frac{2\alpha}{Q_c} \tag{4-14}$$

$$E = \frac{C}{{Q_c}^2} + 4 \tag{4-15}$$

$$G = \sqrt{E^2 - 4D^2} \tag{4-16}$$

$$Q = \sqrt{\frac{E + G}{2D^2}} \tag{4-17}$$

$$K = \frac{\alpha Q}{Q_c} \tag{4-18}$$

$$W = K + \sqrt{K^2 - 1} \tag{4-19}$$

$$F_{01}' = \frac{F_o}{W} \tag{4-20}$$

$$F_{02}' = WF_o \tag{4-21}$$

The result is two band-pass pole pairs having center frequencies of F_{01}' and F_{02}' with identical Q's as defined by equation 4-17. The calculations required by equations 4-13 thru 4-21 should be maintained to over four significant figures after the decimal point to provide adequate accuracy of the results since differences of close numbers are involved. Equations 4-18 and 4-19 are especially critical and should be kept to at least five or six places after the decimal point.

Example 4-1

Given: Band-Pass Filter Requirement
Center Frequency 100Hz
3db Points ±5Hz (95Hz, 105Hz)
25db Minimum ±15Hz (85Hz, 115Hz)

Result: a) Convert to Geometrically Symmetrical band-pass filter requirement (see section 2.2c)[1]

$$F_o = 100 \text{ Hz}$$
$$\text{3db Band-Width} = 10 \text{ Hz}$$
$$\text{25db Band-Width} = 28 \text{ Hz}$$

b) Calculate A_s and select Normalized Low-Pass Filter

$$A_s = \frac{\text{Stop-Band Band-Width}}{\text{3db Band-Width}} = \frac{28}{10} = 2.8$$

Figure 2-6 indicates that a normalized 3 pole Butterworth has more than 25db attenuation at 2.8 radians which satisfies the requirement.

c) Transform Normalized Low-Pass Poles into Band-Pass Pole Pairs

Calculate $Q_c = 10$ (4-10)

using table 7.3, N = 3 Butterworth yields:
real pole $\alpha = 1$, complex pole $\alpha = .500$, $\beta = .866$

Real Pole Transformation

$\alpha = 1$

$Q = 10$ (4-11)

$F_o' = 100\text{Hz}$ (4-12)

Complex Pole Transformation

$\alpha = .500 \quad \beta = .866$

C =	1.00000	(4-13)	D =	.10000	(4-14)
E =	4.01000	(4-15)	G =	4.00501	(4-16)
Q =	20.01875	(4-17)	K =	1.0009375	(4-18)
W =	1.04425	(4-19)	F_{01}' =	95.8Hz	(4-20)
			F_{02}' =	104.4Hz	(4-21)

d) Calculate Component Values

Choose C = .100uf
Select Configuration of Figure 4-2
Choose Unity Gain at F_o (Let A =1)

F_o' = 100Hz
Q = 10
A_o = 200 (4-2)
R_{1a} = 159.2K (4-5)
R_{1b} = 799.8Ω (4-6)
R_2 = 318.3K (4-3)

F_{01}' = 95.8Hz
Q = 20
A_o = 396.3 (4-2)
R_{1a} = 164.6K (4-5)
R_{1b} = 416.4Ω (4-6)
R_2 = 664.5K (4-3)

F_{02}' = 104.4Hz
Q = 20
A_o = 408.1 (4-2)
R_{1a} = 155.5K (4-5)
R_{1b} = 382.1Ω (4-6)
R_2 = 609.8K (4-3)

Figure 4-4 illustrates the circuit configuration and response.

4.3 ALIGNMENT TECHNIQUES

Attempting to build the circuit of Figure 4-4 exactly as shown creates three basic problems:

a) The resistor values calculated by the formulas are not readily available and must be especially ordered from the manufacturer.

b) Frequency tuning errors occur because of component tolerances. For example, capacitor tolerances of 2% and resistor tolerances of 1% could cause a resonant frequency detuning by as much as 3% which would be intolerable in very narrow filters.

c) Amplifier finite open-loop gain and phase shift will cause a frequency detuning. This effect is most pronounced for narrow filters where amplifier open-loop gain requirements are higher and for frequencies in the upper range of the operational amplifier where amplifier gain has dropped off and phase shift has increased.

Although the State-Variable band-pass configuration of Figure 4-3 is less affected by amplifier open loop gain than the single amplifier configuration, amplifier phase shift is still a critical factor for narrow-band filters.

[1] Since this filter is narrow in band-width the specified arithmetically symmetrical requirement could be used to calculate A_S with little error.

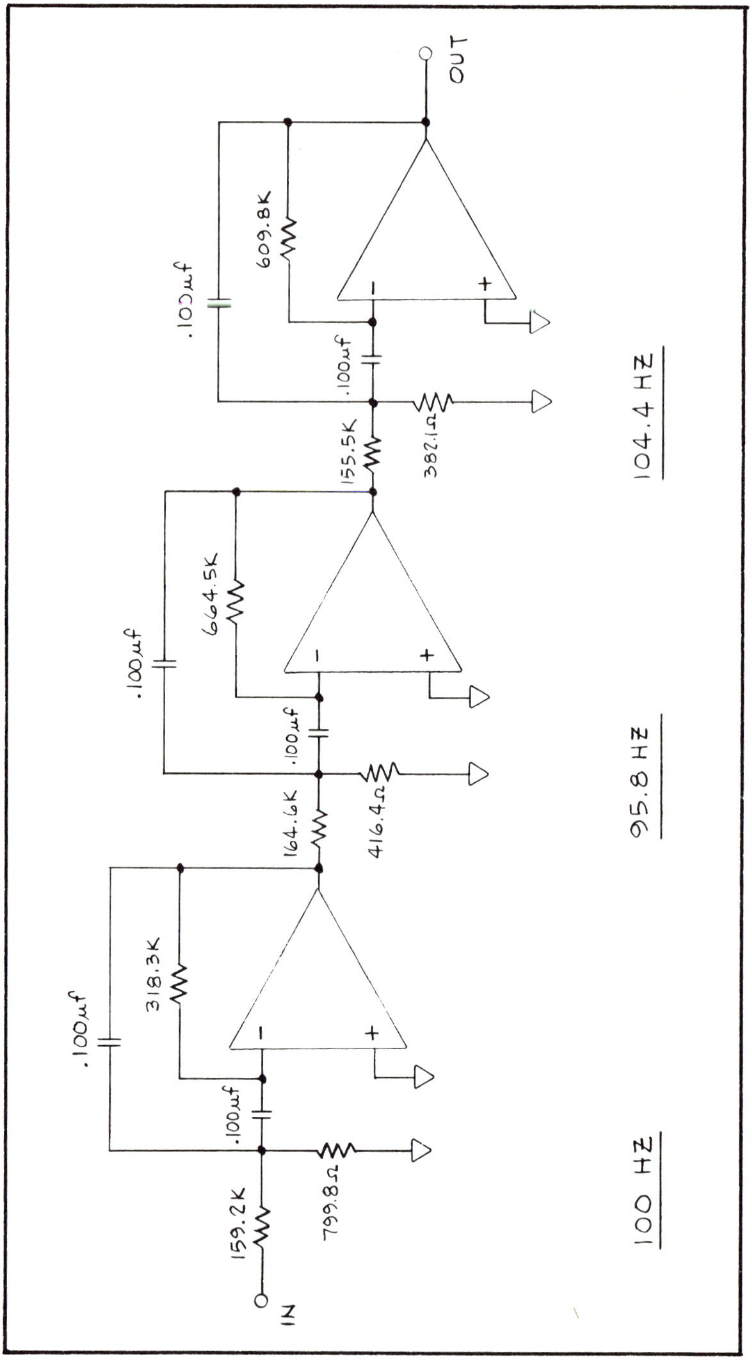

Figure 4-4a
Three-Section Band-Pass
Filter of Example 4-1

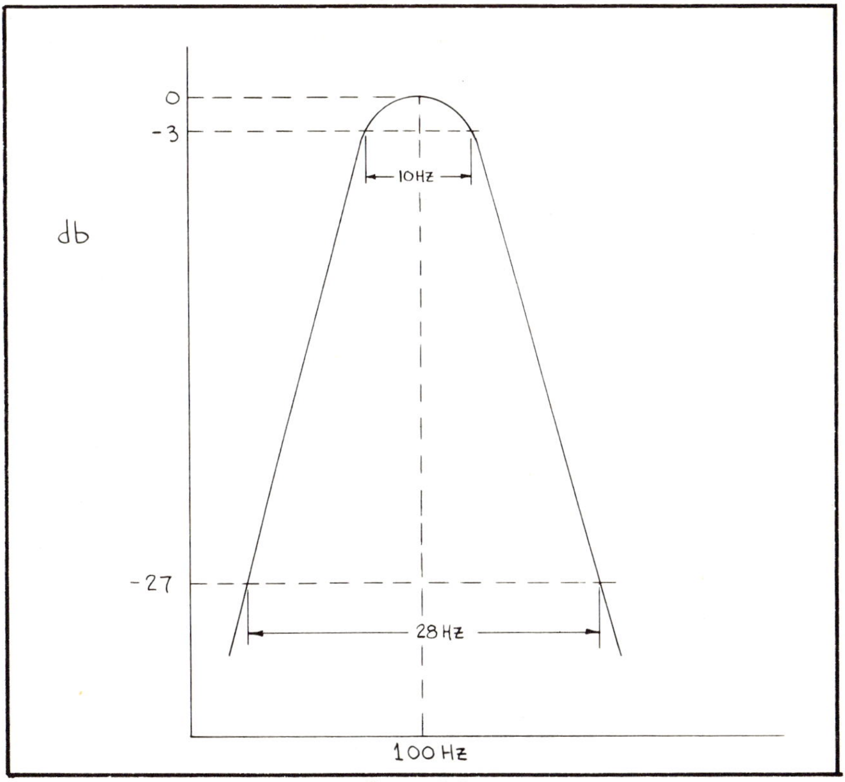

Figure 4-4b
Response of Example 4-1

The result of detuning is a distortion of the frequency response shape. Figure 4-5 shows some of the possible effects of tuning errors in a band-pass filter.

In order to acheive a more precise response some means of adjustment must be provided especially in very narrow filters. The band-pass sections of figures' 4-1 thru 4-3 have a phase shift of exactly 180° at resonance, F_o'. This phase shift provides an excellent criteria for adjustment.

a. Frequency Adjustment

The resonant frequency of the active band-pass sections can be adjusted by varying a particular resistor. In the circuit of Figure 4-1 R_1 will control tuning. Adjusting R_{1a} will tune Figure 4-2. To adjust resonance of the State Variable circuit R_3 of figure 4-3 should be made adjustable.

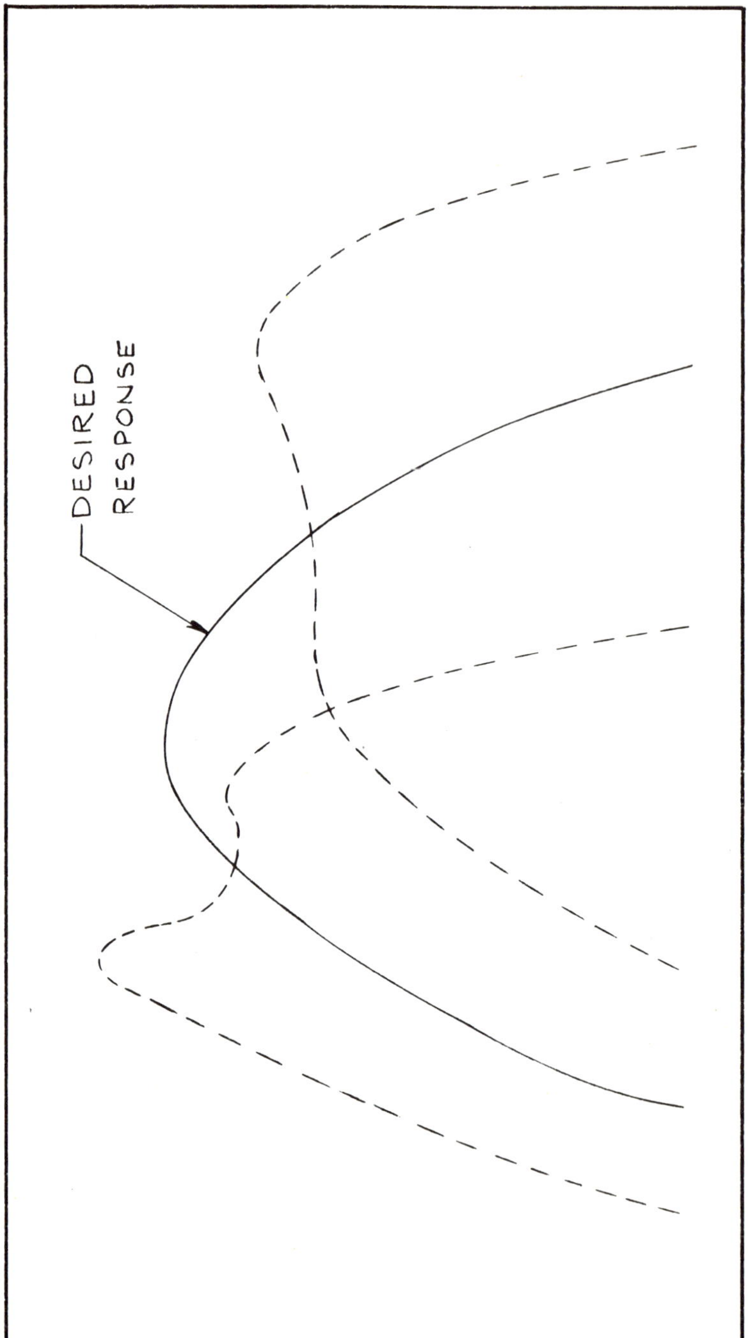

Figure 4-5
Effects of Tuning Errors
in Band-Pass Filters

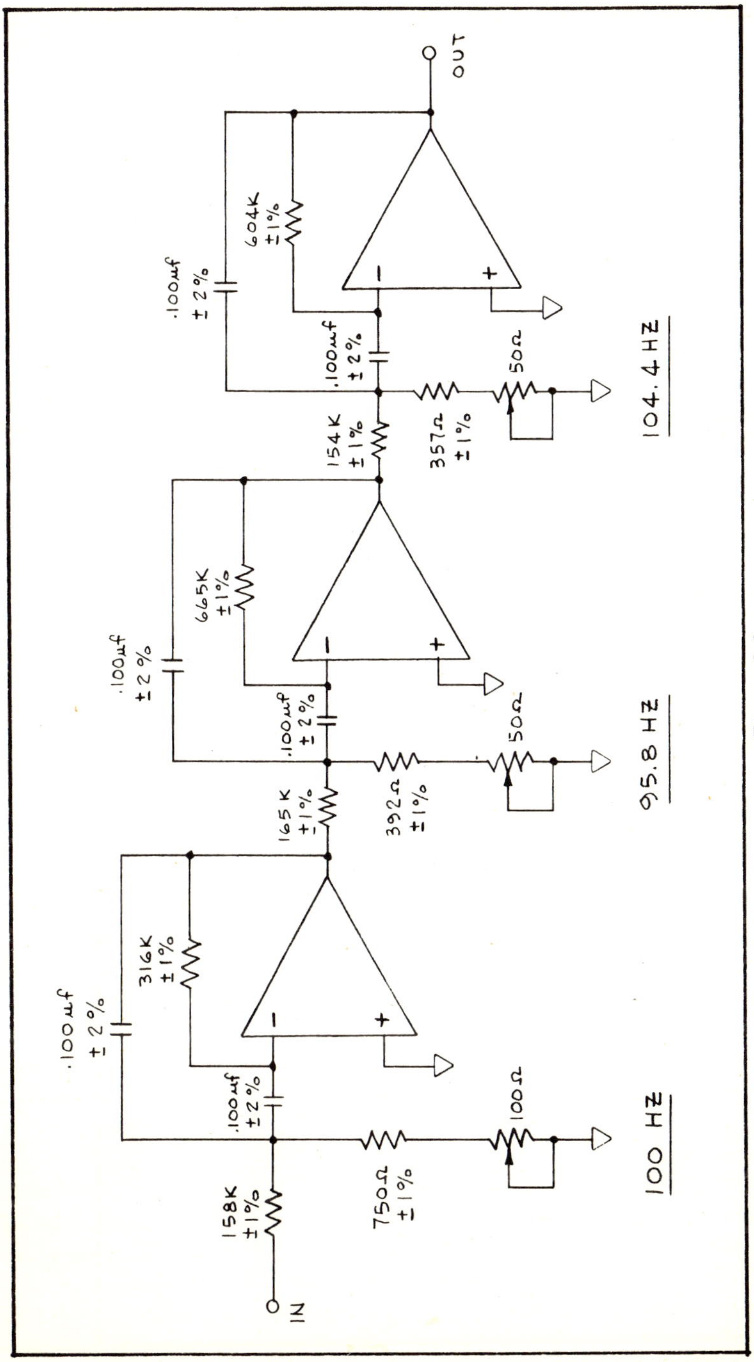

Figure 4-6
Practical Configuration for Example 4-1

If each section has an adjustable center frequency this adjustment can be used to compensate for errors due to tolerances and amplifier detriments. The computed resistance values can then be rounded off to the nearest standard value (see section 6.3). In order to obtain good resolution and stability a fixed precision resistor is placed in series with the control so that the combined resistor value and potentiometer mid-position magnitude approximately equals the calculated value. The resistance adjustment range should be in excess of ± twice (resistance tolerance + capacitor tolerance). Figure 4-6 illustrates the practical realization of the circuit of example 4-1 using these guidelines.

Tuning proceeds in the following manner:

Apply a sine-wave oscillator to the filter input. Proceeding from left to right (filter input towards filter output) monitor the phase shift of each section by connecting the vertical channel of an oscilloscope to section input and the horizontal channel to section output. Set the oscillator for section frequency F_o'. An ellipse or "Lissajous" pattern will be observed. Adjust the tuning resistor until the ellipse collapses to a straight line (at a 135° angle). This should be performed for each section. Care should be taken that the oscilloscope has no internal phase shift over the frequency range of operation. This can be verified by paralleling vertical and horizontal channels across the oscillator and confirming a collapsed ellipse (at a 45° angle).

b. Q Adjustment

Section Q of the circuits of figures' 4-1 and 4-2 is not particularly sensitive to component tolerances or amplifier deficiencies and need not be considered. In any event the filter response is not severely affected by small changes in Q.

In the State-Variable circuit, amplifier phase shift creates an effect called "Q enhancement" where section Q could be increased to a magnitude where the effect on overall response could be significant. This effect can be compensated for by adjustment of section Q as follows:

After adjusting resonance by trimming R_3, adjust R_1 so that the gain between section input and section output at F_o' is the ratio of R_1/R_4 as computed from formulas' 4-7 and 4-9.

CHAPTER 5

Unconventional Active Filters

5.1 NOTCH NETWORKS

A commonly used passive notch filter is the "Twin T". This network provides a deep null at a desired frequency. However the circuit Q which is defined as:

$$Q = \frac{\text{Null Frequency}}{\text{3db Band-Width}} \qquad (5\text{-}1)$$

is only ¼ which results in an extremely poor approximation to a narrow notch.

A technique known as "Boot-Strapping" involves the use of active elements in a feed-back configuration with the Twin-T. This method increases the Q to obtain narrower bandwidths by feeding a portion of the output back to the input. Figure 5-1 illustrates the circuit configuration.

The component values are computed by:

$$R_1 = \frac{1}{2\pi F_o C} \qquad (5\text{-}2)$$

and

$$K = \frac{4Q\text{-}1}{4Q} \qquad (5\text{-}3)$$

R and C may be arbitrarily chosen.

Very high Qs' should be avoided since tuning errors resulting from component tolerances may cause a significant decrease of rejection at the desired notch frequency due to a shifting of the actual notch. About 40 to 60 db rejection could be expected at the notch using 1% resistors and 2% capacitors.

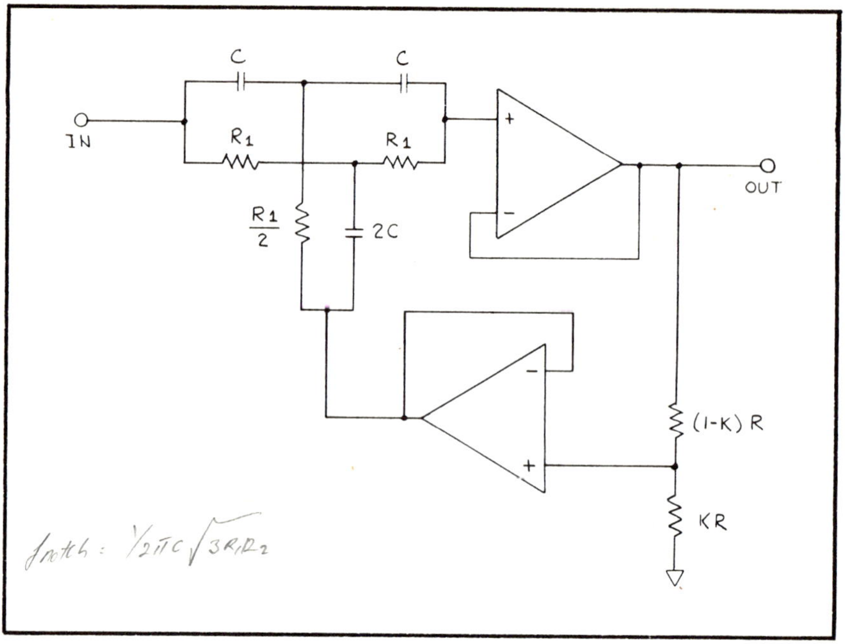

Figure 5-1
Boot-Strapped Twin-T

The attenuation at any band-width, BW_x, can be computed by:

$$DB = 10 \, \text{Log}_{10} \left[1 + \left(\frac{F_o}{Q \cdot BW_x} \right)^2 \right] \tag{5-4}$$

where F_o is the null frequency and Q is defined by equation 5-1. Figure 5-2 shows the response shape.

The impedance of the driving source should be significantly less than the calculated value of R_1.

The function of the unity gain amplifier in the feedback path is to provide a low impedance for the feedback signal. This can be approximated by eliminating the feedback amplifier and directly connecting the feedback signal to the Twin-T provided that R is selected so that (1-k) R is much less than R_1. Figure 5-3 illustrates the simplified configuration.

5.2 DELAY EQUALIZERS AND PHASE-SHIFT NETWORKS

The function of a delay equalizer is to compensate for unwanted

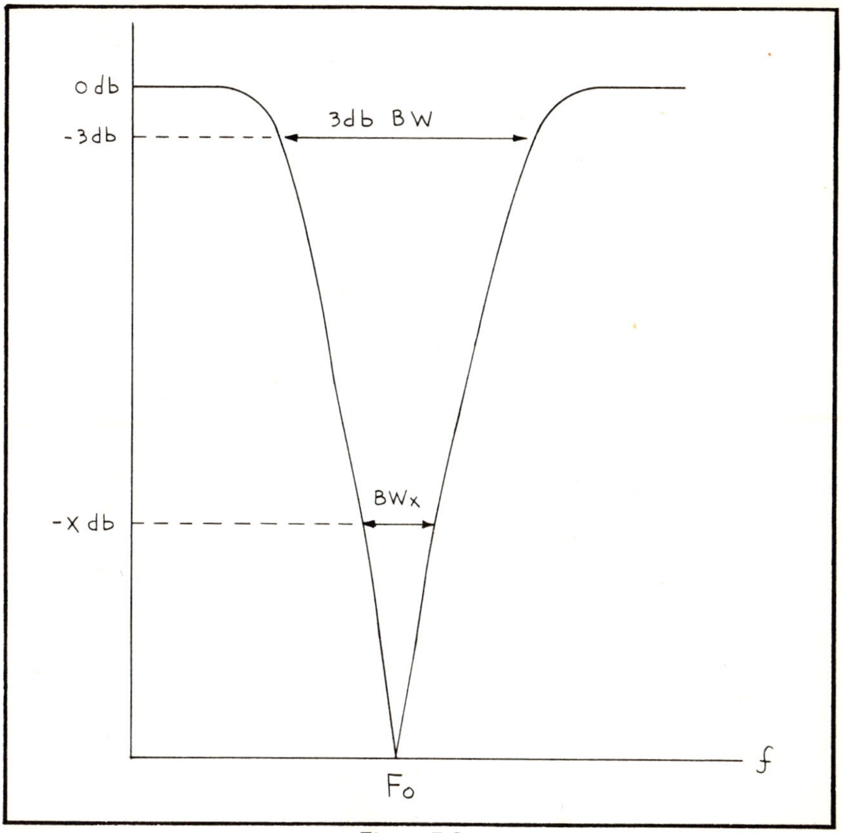

Figure 5-2
Notch Filter Response

delay variations of filters, telephone lines, and other mediums used for transmission of information. This is accomplished by adding delay equalizer sections having the required shape so that the composite delay curve is equalized.

It may be desirable to combine delay equalization and amplitude equalization where the delay distortion of the transmission medium is accompanied by large amplitude ripples. Adjustment capability may be another desired feature so that the equalizer can be adapted to a changed medium.

Phase shift networks are used to introduce a fixed or adjustable amount of phase shift at a desired frequency or over a frequency range.

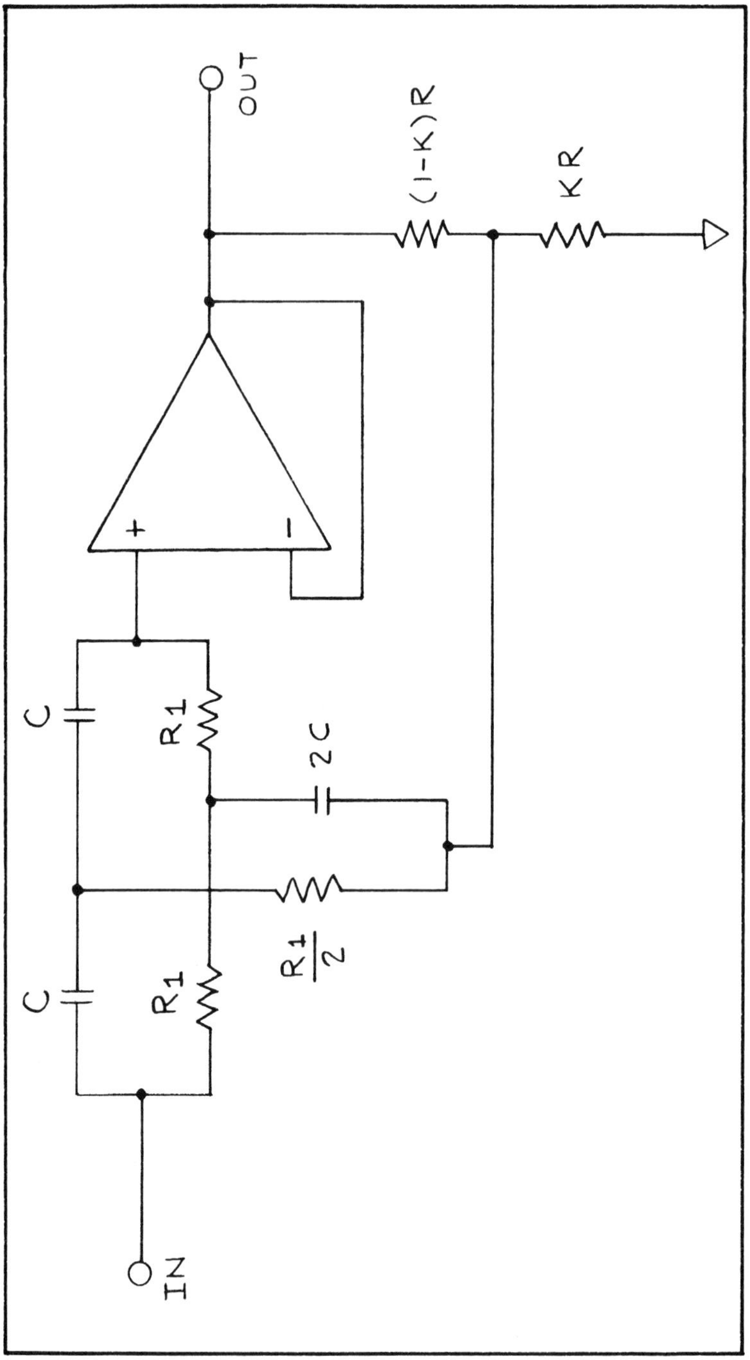

Figure 5-3
Single Amplifier Notch Filter

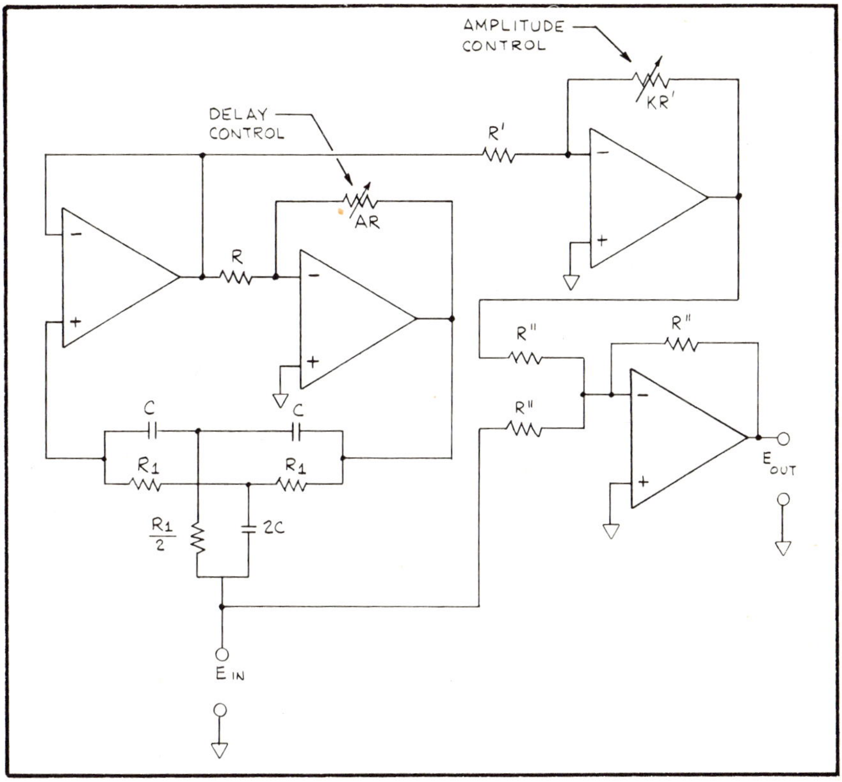

Figure 5-4
Combined Delay and Amplitude Equalizer

a. Adjustable Delay and Amplitude Equalizer

The circuit of figure 5-4 uses the twin-tee of section 5-1 to construct an equalizer which provides both amplitude and delay equalization. The amount of amplitude and delay correction can be adjusted almost independently.

This circuit produces a peaked delay curve having a maximum value in the region of F_o equal to:

$$Tgd = \frac{1 + A}{2\pi F_o} \tag{5-5}$$

where A is determined by the delay control.

The overall circuit gain is unity or Odb at frequencies far removed above and below F_o. However, at F_o, the gain can be greater or less than unity and can be expressed in db as:

$$\text{Amplitude Equalization (db)} = 20 \operatorname{Log}_{10} (K-1) \qquad (5\text{-}6)$$

where K is controlled by the amplitude control.

With K set equal to 2 the circuit becomes "All-Pass". An All-Pass network has unity gain at all frequencies.

The delay and amplitude settings have slight interaction. Delay setting variations do not affect the amount of amplitude equalization at F_o. It does affect the selectivity in the region around F_o. Minor variations of the amplitude control slightly alter the peak delay.

The values of C, R, R′ and R″ may be chosen to be any convenient values. R_1 is calculated using formula 5-2.

Figure 5-5 contains typical delay and amplitude curves.

b. All-Pass Delay Equalizer

This circuit provides a fixed amount of delay equalization while maintaining unity gain over the frequency range of operation. The configuration shown in figure 5-6 uses only two operational amplifiers and may be tuned to correct for the effect of component tolerances and amplifier deficiencies.

After choosing convenient values of C and R compute:

$$R_1 = \frac{1}{2(\pi F_o)^2 \, Tgd \, C} \qquad (5\text{-}7)$$

$$R_2 = \frac{Tgd}{2C} \qquad (5\text{-}8)$$

$$R_3 = \frac{R \, R_2}{4 \, R_1} \qquad (5\text{-}9)$$

where Tgd is the peak delay in seconds, which occurs at F_o.

At resonance the phase shift is 180° between E_{IN} and E′. This provides a criteria for adjustment. By making R_1 variable resonance can be adjusted to the desired F_o using the "Lissajous Pattern" technique discussed in section 4.3a.

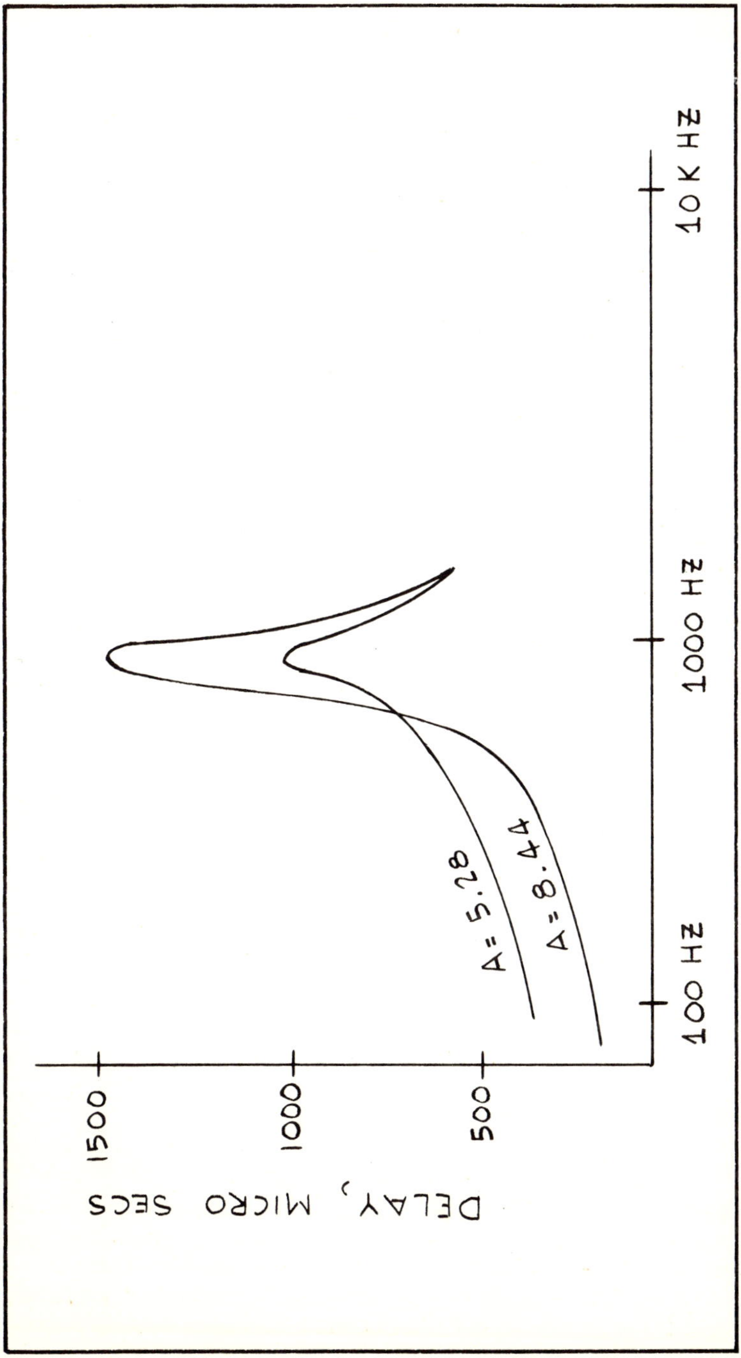

Figure 5-5a
Delay For F_o of 1000Hz,
K of 2 (ALL-PASS)

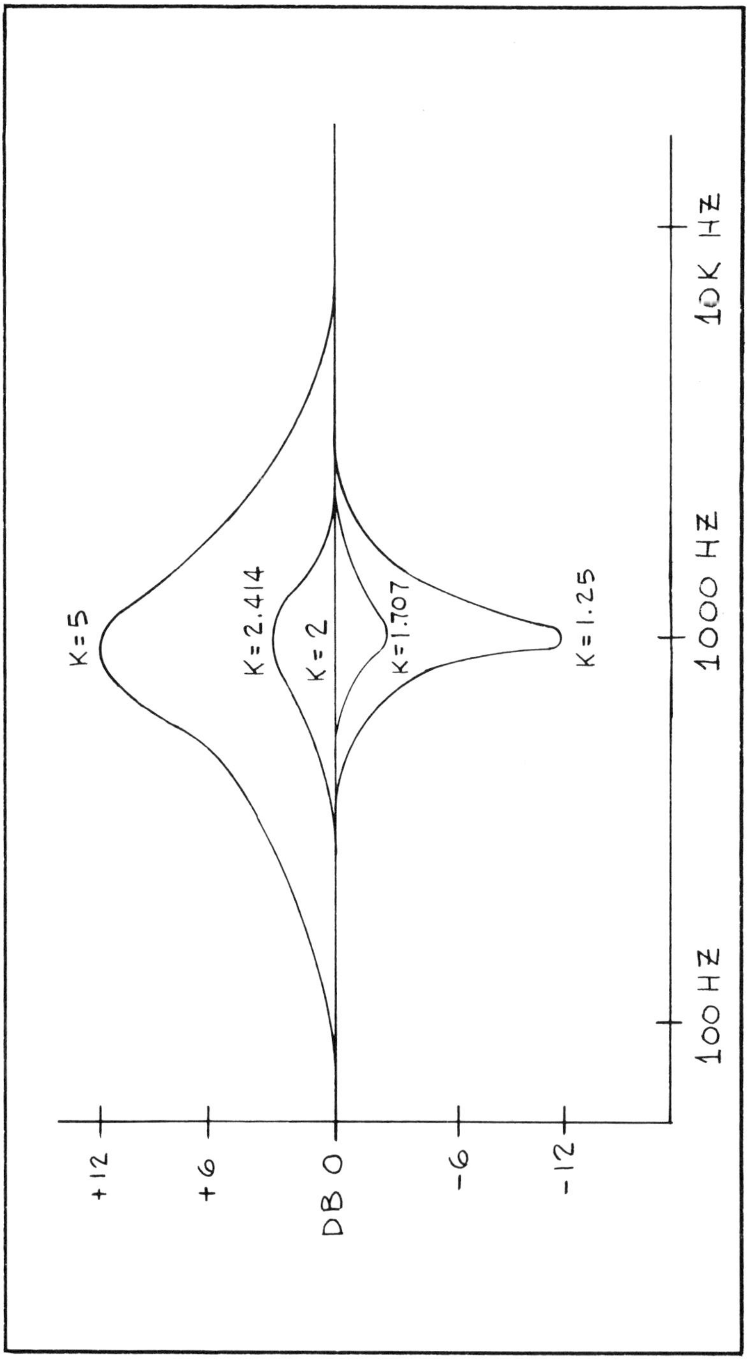

Figure 5-5b
Amplitude for F_o of 1000Hz,
A of 5.28 (Tgd-1000U sec)

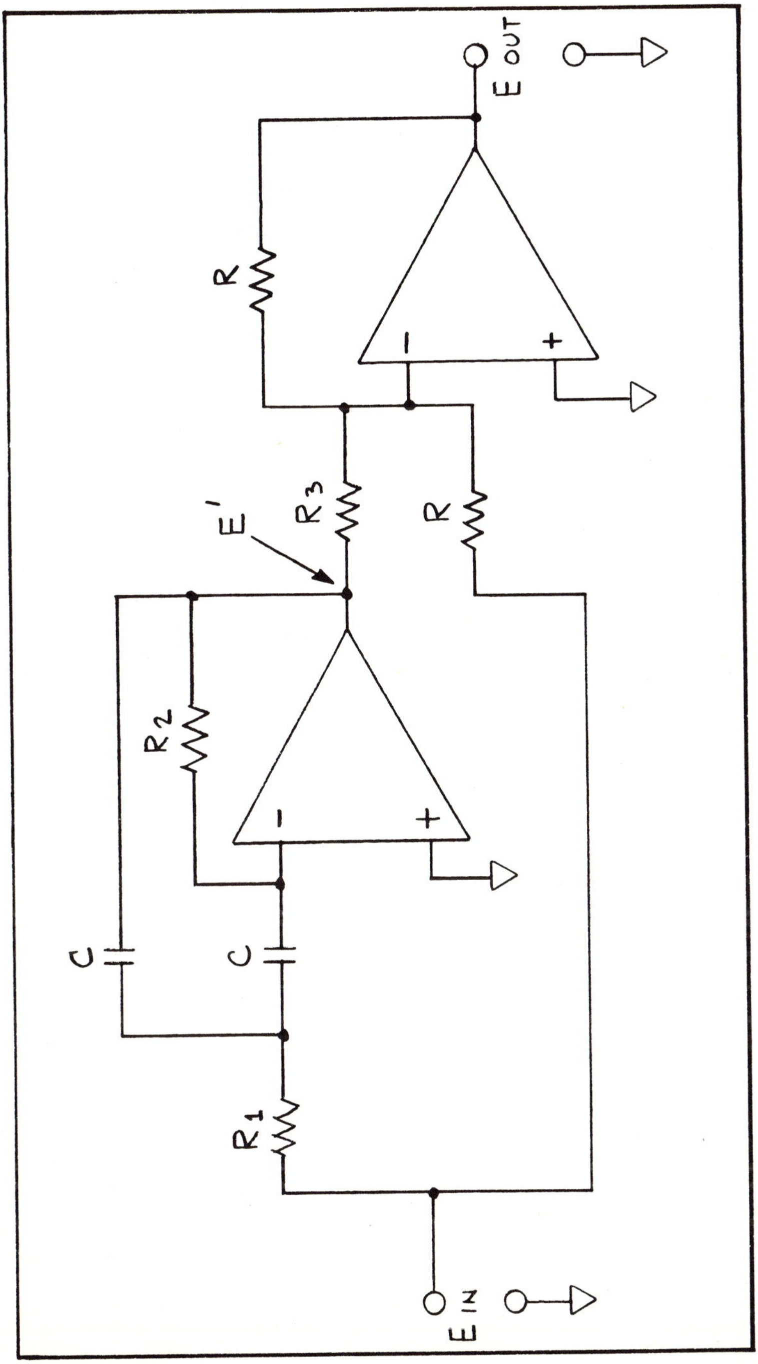

Figure 5-6
All-Pass Delay Equalizer

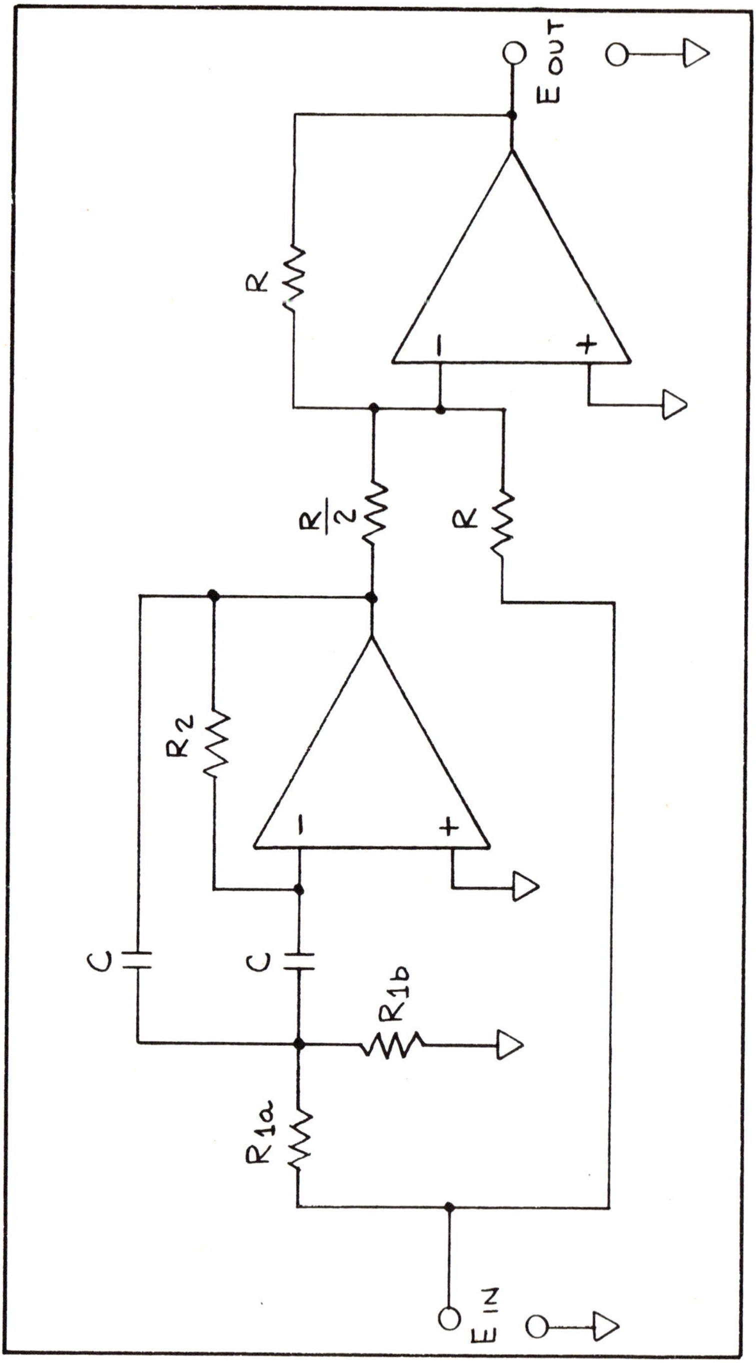

Figure 5-7
Modified All-Pass Delay Equalizer

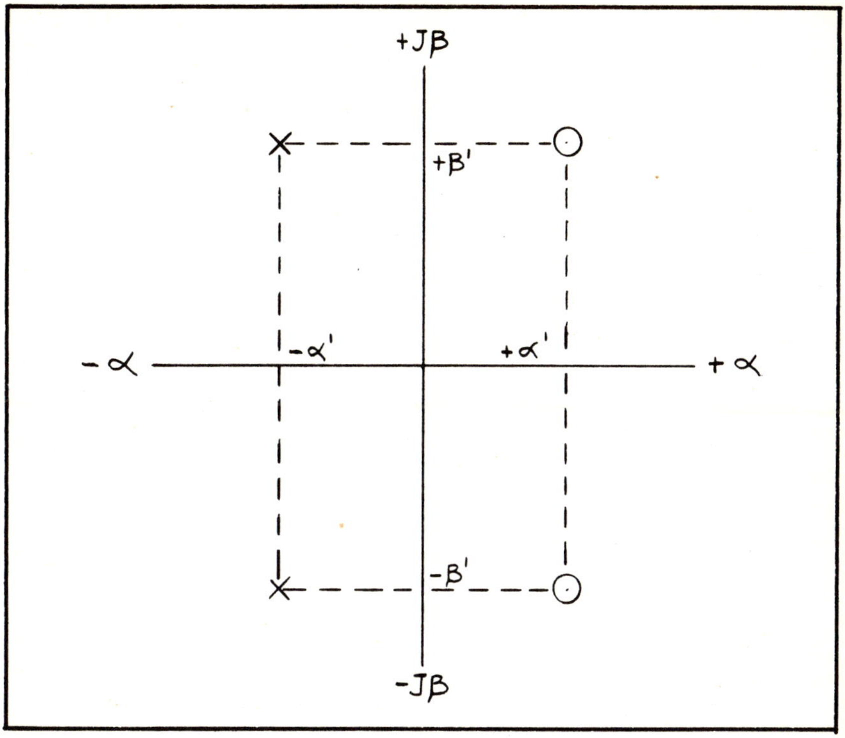

Figure 5-8
Pole-Zero Pattern for
All-Pass Delay Equalizer

Although the overall circuit gain is unity over the frequency range, the gain between E_{IN} and E' behaves in a band-pass fashion and reaches a maximum value at F_o which can be calculated by:

$$A' = \frac{(\pi F_o \, Tgd)^2}{2} \qquad (5\text{-}10)$$

Clearly, Large F_o Tgd products result in Large gains which may cause amplifier clipping. A solution for these cases is to replace R_1 of figure 5-6 with a voltage divider. The modified circuit is shown in figure 5-7 where:

$$R_{1a} = \frac{R_2}{2} \qquad (5\text{-}11)$$

$$R_{1b} = \frac{R_1 \, R_2}{R_2 - 2R_1} \qquad (5\text{-}12)$$

The transfer function of the All-Pass Delay Equalizer, as well as the adjustable Equalizer when set for the All-Pass condition, consists of Two Poles and Two Zeros. These roots are located in the corners of a rectangle having sides $2\alpha'$ and $2\beta'$ as illustrated in figure 5-8. F_o and Tgd are related to α' and β' by:

$$F_o = \frac{\sqrt{\alpha'^2 + \beta'^2}}{2\pi} \qquad (5\text{-}13)$$

$$Tgd = \frac{2}{\alpha'} \qquad (5\text{-}14)$$

c. Adjustable Phase-Shift Network

The circuit of figure 5-9 provides a phase shift which can be adjusted from 0° to -180° while maintaining All-Pass properties. The midrange setting of -90° occurs at F_o where:

$$F_o = \frac{1}{2\pi\, RC} \qquad (5\text{-}15)$$

C and R′ may be arbitrarily selected. To calculate R for a desired phase shift of $-\phi°$ at F′, use the following relationship:

$$R = \frac{\tan(\phi/2)}{2\pi F' C} \qquad (5\text{-}16)$$

Since R is a function of the Tangent of the angle $\phi/2$, angles near 0° or -180° result in extreme values of R (o ohms or infinite ohms) so the practical range is generally restricted to between -45° and -135°.

5.3 STATE-VARIABLE ELLIPTIC FUNCTION FILTERS

Where high precision Elliptic Function filters are required, the State-Variable configuration offers adjustment capability whereas the approach of chapter 3 does not. In addition, the circuit is more tolerant of component tolerances and amplifier deficiencies. These features are acheived at the cost of some additional circuit complexity.

Normalized Low-Pass Elliptic Function poles and zeros are provided in Table 7.2 alongside the tabulated values. By suitable denormalization, these poles and zeros may be directly implemented using a cascade of State-Variable sections for Low-Pass filters. High-Pass and Band-Pass filters first require a Low-Pass to High-Pass or Low-Pass to

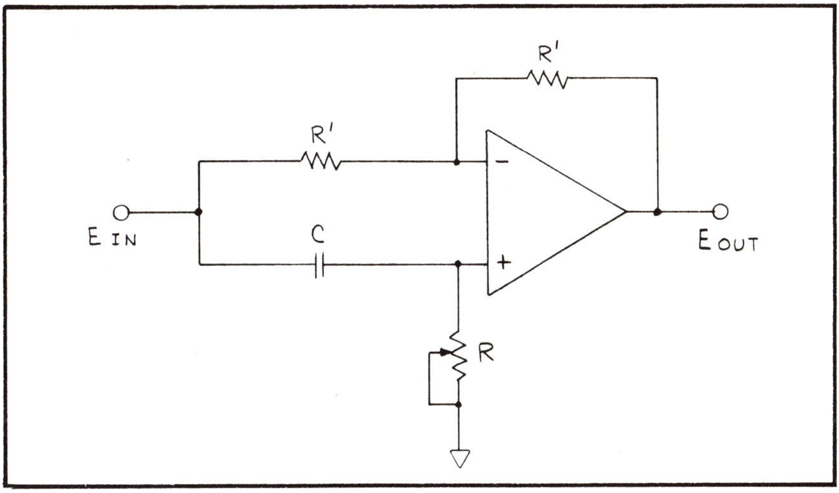

Figure 5-9
Adjustable Phase-Shift Network

Band-Pass transformation on the poles and zeros. The same basic configuration can then be used.

Adjustment of section resonant frequency, Q, and "Notch" or Zero frequency can be made independently. The result is a highly stable filter whose response is in good agreement with theoretical predictions.

a. State-Variable Building Block

A State-Variable Band-Pass configuration was presented in section 4-1b. This circuit provided a pair of band-pass poles. The circuit of figure 5-10 can realize a pair of poles and zeros. By connecting either strap 1 or 2, the zero can be placed above or below the section resonant frequency. The basic design parameters are F_0', Q and F_∞ where F_0' is section resonant frequency in Hz, Q is section selectivity factor and F_∞ is section infinite rejection frequency (Notch) in Hz.

To calculate the component values first compute:

$$a = \frac{2\pi F_0'}{Q} \tag{5-17}$$

$$b = (2\pi F_0')^2 \tag{5-18}$$

$$d = (2\pi F_\infty)^2 \tag{5-19}$$

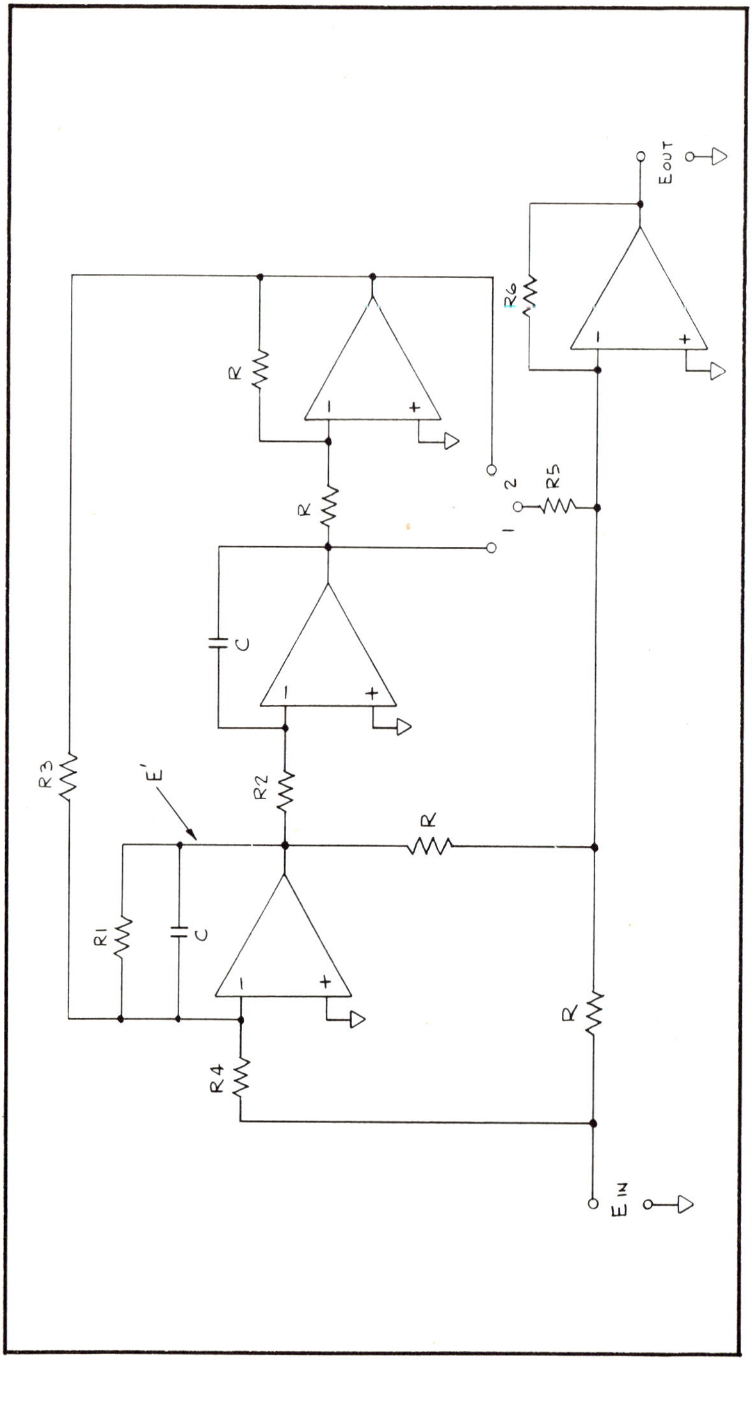

Figure 5-10
State-Variable Building Block

Select convenient values for R and C. The remaining resistor values are:

$$R_1 = R_4 = \frac{1}{a\,C} \tag{5-20}$$

$$R_2 = R_3 = \frac{1}{\sqrt{b}\;C} \tag{5-21}$$

$$R_5 = \frac{a\sqrt{b}\,R}{|b\text{-}d|} \tag{5-22}$$

$$\text{for } F_\infty > F_0' \qquad R_6 = \frac{R}{1 + \frac{|b\text{-}d|}{b}} \tag{5-23}$$

$$\text{for } F_\infty < F_0' \qquad R_6 = R \tag{5-24}$$

The computed values provide unity section gain. The gain may be raised or lowered by proportionally changing R_6. The values of F_0', Q and F_∞ for each section are derived from the Elliptic-Function poles and zeros of Table 7.2 transformed in accordance with the remaining sections of this chapter.

The section can be adjusted by implementing the following steps in the following sequence.

1. Resonant Frequency

By making R_3 adjustable, the section resonant frequency can be tuned to the required F_o'. The "Lissajous Pattern" method of section 4.3 can be used in using E′ as the output to the scope horizontal channel. The resistance adjustment range should be in excess of ± twice (resistance tolerance and capacitor tolerance).

2. Q Adjustment

Amplifier phase shift creates a "Q enhancement" effect where section Q is increased. This effect also increases the section gain at E′ above unity. To compensate for "Q enhancement" reduce the value of R_1 until the gain at E′ is unity at F_o'. This will closely provide the design Q.

An alternate technique is to reduce R_1 until the 3db bandwidth at E′ is in accordance with formula 4-10. However, although more accurate, this measure is certainly slower than a unity gain adjustment.

3. Notch Frequency Tuning

Notch frequency tuning usually isn't required since adjustment of F_0' also brings in F_∞. However, if an independent adjustment of F_∞ is desired, R_5 will provide the frequency tuning. The notch frequency is measured by determining the frequency where E_{out} is nulled.

If the notch is above F_o' make connection #1 of figure 5-10. For a notch below F_o' make connection #2.

b. Low-Pass Design

After selecting an Elliptic Function Low-Pass filter from Table 7.2 which satisfies the steepness requirement, A_s, (see section 3.2a and example 3-3). The tabulated poles and zeros must be converted into the F_0', Q and F_∞ for each state-variable section.

Using the tabulated section co-ordinates A, B and Zero where:

A = Real part of pole position

B = Imaginary part of pole position

Zero = Position of Zero

compute:

$$F_0' = F_c \sqrt{A^2 + B^2} \tag{5-25}$$

$$Q = \frac{\sqrt{A^2 + B^2}}{2A} \tag{5-26}$$

$$F_\infty = F_c \times \text{Zero} \tag{5-27}$$

F_c is the desired filter cutoff frequency. The calculated F_o', Q and F_∞ for each section can then be implemented using the circuit of section 5.3a.

The last filter section requires realization of a real pole, A. This real pole can be introduced in the manner of figure 3-4 by adding an RC section where R is optional and

$$C = \frac{1}{2\pi F_c AR} \tag{5-28}$$

This real pole can also be satisfied by introducing a capacitor across R_6 of the last section having the value:

$$C = \frac{1}{2\pi F_c AR_6} \tag{5-29}$$

A state-variable approach is used in the following example:

Example 5-1

Given: Low-Pass Filter Requirement
.5db Maximum Ripple Below 1000Hz
20db Minimum Rejection Above 1560Hz

Result:

a. Compute $A_s = \frac{1560}{1000} = 1.56$

b. Select Filter from table 7.2

$N = 3$
$R_{db} = .28db$
$\omega_s = 1.556$
$A_{db} = 20.58db$

c. Expected Response

.28db at 1000Hz
20.58db at 1556Hz

d. Compute Section F_o', Q, F_∞

$A = .2643$
$B = 1.100$ From Table 7.2
$Zero = 1.742$
$F_c = 1000$ Hz
$F_o' = 1131$ Hz (5-25)
$Q = 2.14$ (5-26)
$F_\infty = 1742$ Hz (5-27)

e. Calculate Section Values
(choose R = 10K C = .01uf)

$a = 3,321$ (5-17)
$b = 50.50 \times 10^6$ (5-18)
$d = 119.8 \times 10^6$ (5-19)
$R_1 = R_4 = 30,110\ \Omega$ (5-20)
$R_2 = R_3 = 14,072\ \Omega$ (5-21)
$R_5 = 3,405\ \Omega$ (5-22)
$R_6 = 4,215\ \Omega$ (5-23)

f. Introduce Real Pole
(Parallel capacitor across R_6)

$A = .9142$ (from table 7.2)
$C = .0412uf$ (5-29)

Figure 5-11 illustrates the final circuit. Note that the resistor values have been modified to allow use of 1% precision resistors and to provide adjustment.

c. High-Pass Filters

In order to design an Elliptic Function State-Variable High-Pass Filter the poles and zeros must first undergo a Low-Pass to High-Pass transformation. The transformed poles and zeros are then converted into the F_o', Q and F_∞ for each section.

To transform a Normalized complex Low-Pass pole-pair having a real part A and an imaginary part B, as tabulated in table 7.2, compute:

$$A_{hp} = \frac{A}{A^2 + B^2} \tag{5-30}$$

$$B_{hp} = \frac{B}{A^2 + B^2} \tag{5-31}$$

resulting in a complex high-pass pole-pair with a real part A_{hp} and an imaginary part B_{hp}.

To transform normalized Low-Pass Zeros into normalized High-Pass Zeros, merely reciprocate the Zeros in table 7.2.

$$Zero_{hp} = \frac{1}{Zero} \tag{5-32}$$

The transformed poles and zeros can then be converted into the design parameters of the circuit of figure 5-10 where:

$$F_0' = F_c \sqrt{A_{hp}^2 + B_{hp}^2} \tag{5-33}$$

$$Q = \frac{\sqrt{A_{hp}^2 + B_{hp}^2}}{2A_{hp}} \tag{5-34}$$

$$F_\infty = F_c \times Zero_{hp} \tag{5-35}$$

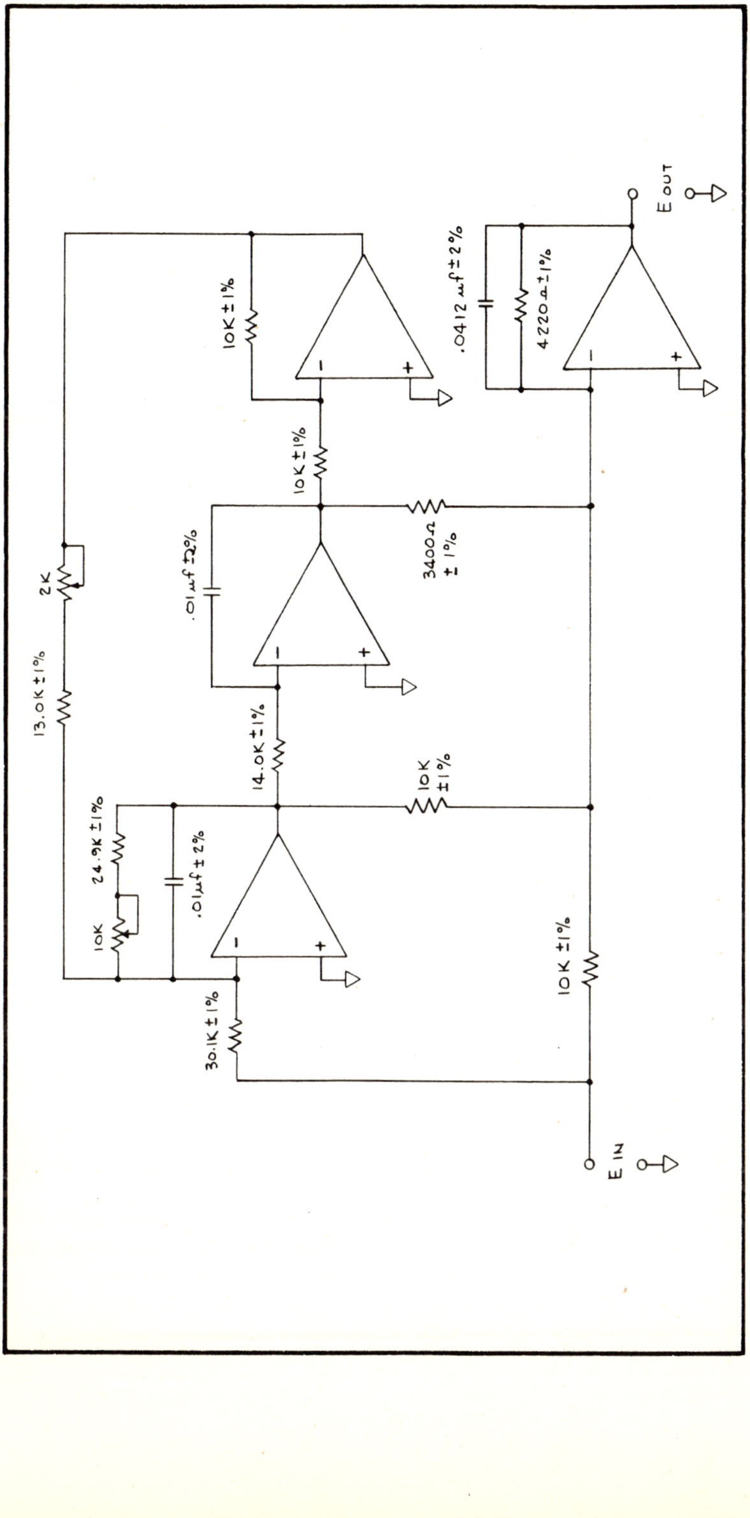

Figure 5-11
State Variable Low-Pass Filter of Example 5-1

F_c is the design cutoff frequency. Since the notch frequency, F_∞, occurs below F_o', connection #2 of figure 5-10 must be made. Equations 5-17 thru 5-24 will provide the component values.

The real pole, A, associated with the last section, can be transformed into a normalized high-pass real pole by reciprocating.

$$A_{hp} = \frac{1}{A} \tag{5-36}$$

This normalized high-pass real pole can be denormalized and realized as an RC section identical to the last section of figure 3-7c where R is optional and C can be calculated by:

$$C = \frac{1}{2\pi\, F_c A_{hp} R} \tag{5-37}$$

Design of a high-pass Elliptic-Function filter using the State-Variable approach is demonstrated in the following example.

Example 5-2

Given: High-Pass Filter Requirement
.5db Maximum Ripple Above 1000Hz
20db Minimum Rejection Below 640Hz

Result: a. Compute $A_s = \dfrac{1000}{640} = 1.562$

b. Select Filter from Table 7.2

$N = 3$
$R_{db} = .28\text{db}$
$\omega_s = 1.556$
$A_{db} = 20.58\text{db}$

c. Expected Response

.28db at 1000Hz

20.58db at $\dfrac{1000}{1.556}$ or 642.7Hz

d. Transform normalized complex Low-Pass Pole-Pair to High-Pass Pole-Pair and transform Zero to High-Pass Zero.

$A = .2643$
$B = 1.100$ From Table 7.2

$$\begin{aligned} \text{Zero} &= 1.742 \\ A_{hp} &= .2065 && (5\text{-}30) \\ B_{hp} &= .8595 && (5\text{-}31) \\ \text{Zero}_{hp} &= .5741 && (5\text{-}32) \end{aligned}$$

e. Compute Section F_o', Q, F_∞

$$\begin{aligned} F_c &= 1000 \text{ Hz} \\ F_o' &= 884 \text{ Hz} && (5\text{-}33) \\ Q &= 2.14 && (5\text{-}34) \\ F_\infty &= 574 \text{ Hz} && (5\text{-}35) \end{aligned}$$

f. Calculate Section Values
(Choose R = 10K C = .01uf)

$$\begin{aligned} a &= 2{,}595 && (5\text{-}17) \\ b &= 30.85 \times 10^6 && (5\text{-}18) \\ d &= 13.01 \times 10^6 && (5\text{-}19) \\ R_1 = R_4 &= 38{,}536\ \Omega && (5\text{-}20) \\ R_2 = R_3 &= 18{,}004\ \Omega && (5\text{-}21) \\ R_5 &= 8{,}079\ \Omega && (5\text{-}22) \\ R_6 &= 10\text{ K} && (5\text{-}24) \end{aligned}$$

g. Calculate Real Pole
(Choose R = 10K)

$$\begin{aligned} A &= .9142 && \text{(from table 7.2)} \\ A_{hp} &= 1.0939 && (5\text{-}36) \\ C &= .0145 \text{ uf} && (5\text{-}37) \end{aligned}$$

The final circuit is shown in figure 5-12.

d. Band-Pass Filters

Elliptic Function Band-Pass Filters are an extremely powerful filter type. A steeper pass-band to stop-band transition than the all-pole filter type of section 4 can be acheived at the expense of some additional circuit complexity and return lobes in the stop-band. For most filtering requirements, the all-pole type is recommended. Where a steeper rate of roll-off is required the Elliptic Function Band-Pass approach could be used advantageously. If the ratio of upper cutoff frequency to lower cutoff frequency is in excess of 1.5, a Low-Pass High-Pass approach should be used as discussed in section 3.3.

To select a normalized Elliptic Function Low-Pass filter, first convert the band-pass filter requirement into a requirement having Geometric Symmetry, using the method of section 2.2C. Calculate a band-pass steepness factor A_s where:

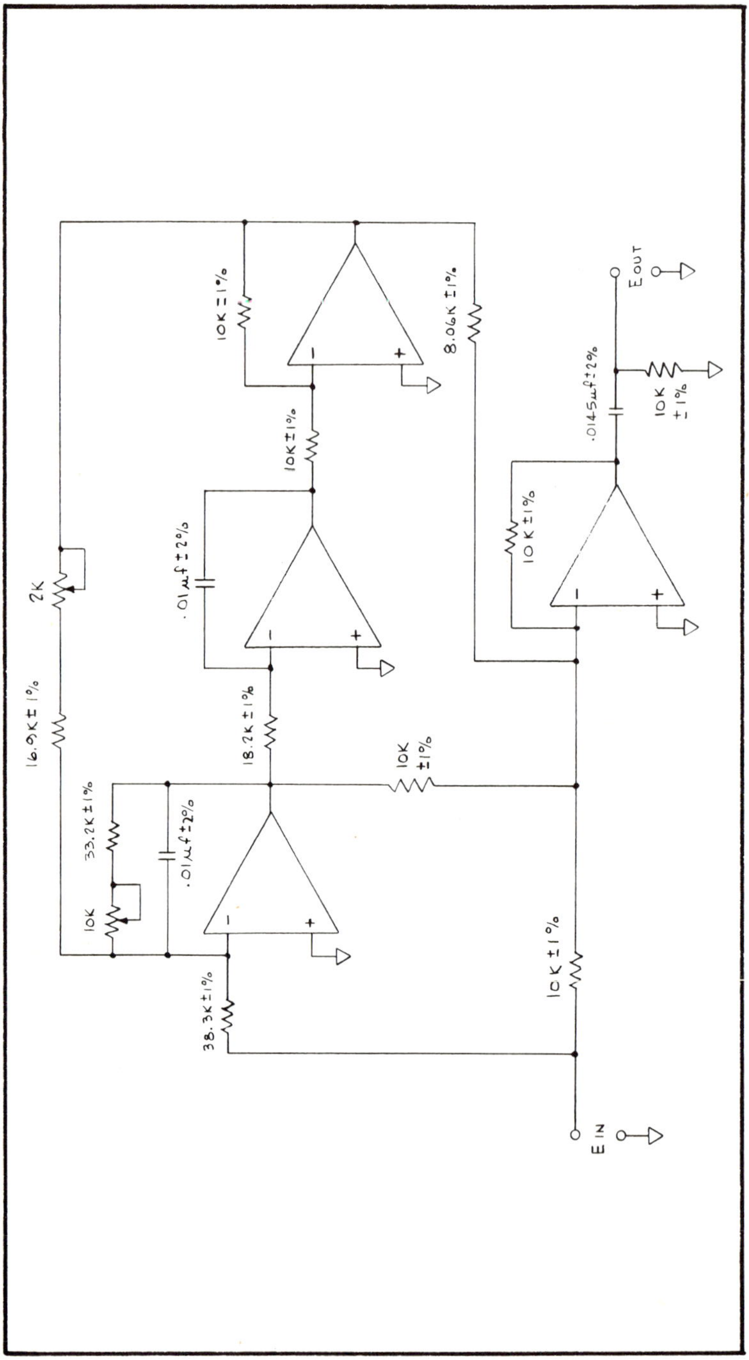

Figure 5-12
State-Variable High-Pass Filter of Example 5-2

$$A_s = \frac{\text{Stop-Band Band-Width}}{\text{Pass-Band Band-Width}} \tag{5-38}$$

A Low-Pass filter is then selected from table 7.2 that satisfies the following three requirements:

1. ω_s should not exceed A_s.
2. A_{db} should satisfy the minimum stop-band rejection required.
3. R_{db} should be less than the maximum allowable pass-band ripple.

A Low-Pass to Band-Pass transformation must be performed on the Low-Pass poles and zeros of each section of the selected filter. Each complex pole-pair is transformed into two complex pole-pairs. Similarly each zero-pair results in two pairs of zeros. The single real pole is transformed into a complex pole-pair.

The band-pass poles and zeros can be defined in terms of F_o', Q and F_∞ which are directly realizable using the circuit of figure 5-10. The complex pole-pair without the associated zeros, which was derived from the real poles, can be obtained by using one of the all-pole band-pass circuits of section 4-1.

The following equations transform each normalized Low-Pass section into two sets of band-pass design parameters, using the A, B and Zero values from table 7.2 along with F_o and Q_c where F_o is the filter geometric center frequency and

$$Q_c = \frac{F_o}{F_{bw}} \tag{5-39}$$

F_{bw} is the filter pass-band bandwidth.

$$C = A^2 + B^2 \tag{5-40}$$

$$D = \frac{2A}{Q_c} \tag{5-41}$$

$$E = \frac{C}{Q_c^2} + 4 \tag{5-42}$$

$$G = \sqrt{E^2 - 4D^2} \tag{5-43}$$

$$Q = \sqrt{\frac{E + G}{2D^2}} \tag{5-44}$$

$$K = \frac{AQ}{Q_c} \tag{5-45}$$

$$W = K + \sqrt{K^2 - 1} \tag{5-46}$$

$$H = \frac{(\text{Zero})^2}{2Q_c^{\,2}} + 1 \tag{5-47}$$

$$Z = \sqrt{H + \sqrt{H^2 - 1}} \tag{5-48}$$

The design parameters for the resulting two bandpass sections are:

SECTION 1

$$F_o' = \frac{F_o}{W} \tag{5-49}$$

$$F_\infty = \frac{F_o}{Z} \tag{5-50}$$

SECTION 2

$$F_o' = WF_o \tag{5-51}$$

$$F_\infty = ZF_o \tag{5-52}$$

Both sections have identical Qs′ as determined by equation 5-42.

Using the State-Variable circuit of Figure 5-10, section 1 requires strap 2 whereas section 2 requires strap 1. Design equations 5-17 thru 5-22 are used for component values.

The last section of the chosen filter from table 7.2 has a real pole, A, which transforms into a single complex pole-pair with the following parameters:

$$F_o' = F_o \tag{5-53}$$

$$Q = \frac{Q_c}{A} \tag{5-54}$$

A single pole-pair configuration can then be chosen from section 4-1.

The following example demonstrates the design of an Elliptic Function Band-Pass filter.

Example 5-3

Given: Band-Pass Filter Requirement
Center Frequency 500 Hz
.5db Maximum attenuation ±50Hz (450Hz, 550Hz)
30db Minimum attenuation ±150Hz (350Hz, 650Hz)

Result:

a. Convert to Geometrically Symmetrical band-pass filter requirement (see section 2.2c)

$F_o = \sqrt{450x550} = 497.5Hz$
.5db Band-Width = 100Hz
30db Band-Width = 269Hz

b. Calculate A_s and select Normalized Low-Pass Filter from table 7.2

$$A_s = \frac{\text{Stop-Band Band-Width}}{\text{Pass-Band Band-Width}} = \frac{269}{100} = 2.69$$

Select:

$N = 3$
$R_{db} = .1db$
$\omega_s = 2.559$
$A_{db} = 31.13$ db

c. Transform Normalized Low-Pass poles and zeros into the Band-Pass Parameters & Compute Component Values

Calculate $Q_c = 4.98$ (5-39)

Transform each normalized complex Low-Pass pole-pair and zero-pair into the design parameters for two band-pass sections.

A = .4249
B = 1.216 (from table 7.2)
Zero = 2.926

C = 1.65920 (5-40)
D = .17064 (5-41)
E = 4.06690 (5-42)
G = 4.05256 (5-43)
Q = 11.808 (5-44)
K = 1.0074737 (5-45)
W = 1.12996 (5-46)
H = 1.17261 (5-47)
Z = 1.33603 (5-48)

The resulting parameters are:

SECTION 1

$F_o' = 440.3Hz$ (5-49)
$Q = 11.8$ (5-44)
$F_\infty = 372.4Hz$ (5-50)

SECTION 2

$F_o' = 562.2Hz$ (5-51)
$Q = 11.8$ (5-44)
$F_\infty = 664.7Hz$ (5-52)

Calculate section Values
(Choose R = 10K C = .047uf)

SECTION 1

$a =$	234.4	(5-17)
$b =$	7.653×10^6	(5-18)
$d =$	5.475×10^6	(5-19)
$R_1 = R_4 =$	90,770 Ω	(5-20)
$R_2 = R_3 =$	7,691 Ω	(5-21)
$R_5 =$	2,977 Ω	(5-22)
$R_6 =$	10 KΩ	(5-24)

SECTION 2

$a =$	299.3	(5-17)
$b =$	12.48×10^6	(5-18)
$d =$	17.44×10^6	(5-19)
$R_1 = R_4 =$	71,090 Ω	(5-20)
$R_2 = R_3 =$	6,023 Ω	(5-21)
$R_5 =$	2,132 Ω	(5-22)
$R_6 =$	7,156 Ω	(5-23)

Transform real pole into single Band-Pass Pole-Pair

$A =$	1.054	(table 7.2)
$F_o' =$	497.5 Hz	(5-53)
$Q =$	4.73	(5-54)

Using circuit of figure 4-2
(Choose C = .047uf, Gain = 1)

SECTION 3

$A_o = 2Q^2 =$	44.7	(4-2)
$R_2 =$	64,390 Ω	(4-3)
$R_{1a} =$	32,160 Ω	(4-5)
$R_{1b} =$	736 Ω	(4-6)

Figure 5-13 shows the final circuit and response.

The series of calculations demonstrated in example 5-3 may become quite tedious for higher order Low-Pass filters. Use of a programmable desk calculator or a computer may be advantageous.

5.4 ACTIVE ALL-PASS DELAY LINES

The Maximally-Flay Delay family of Low-Pass filters features a constant delay. For N greater than 3, this flat behavior extends

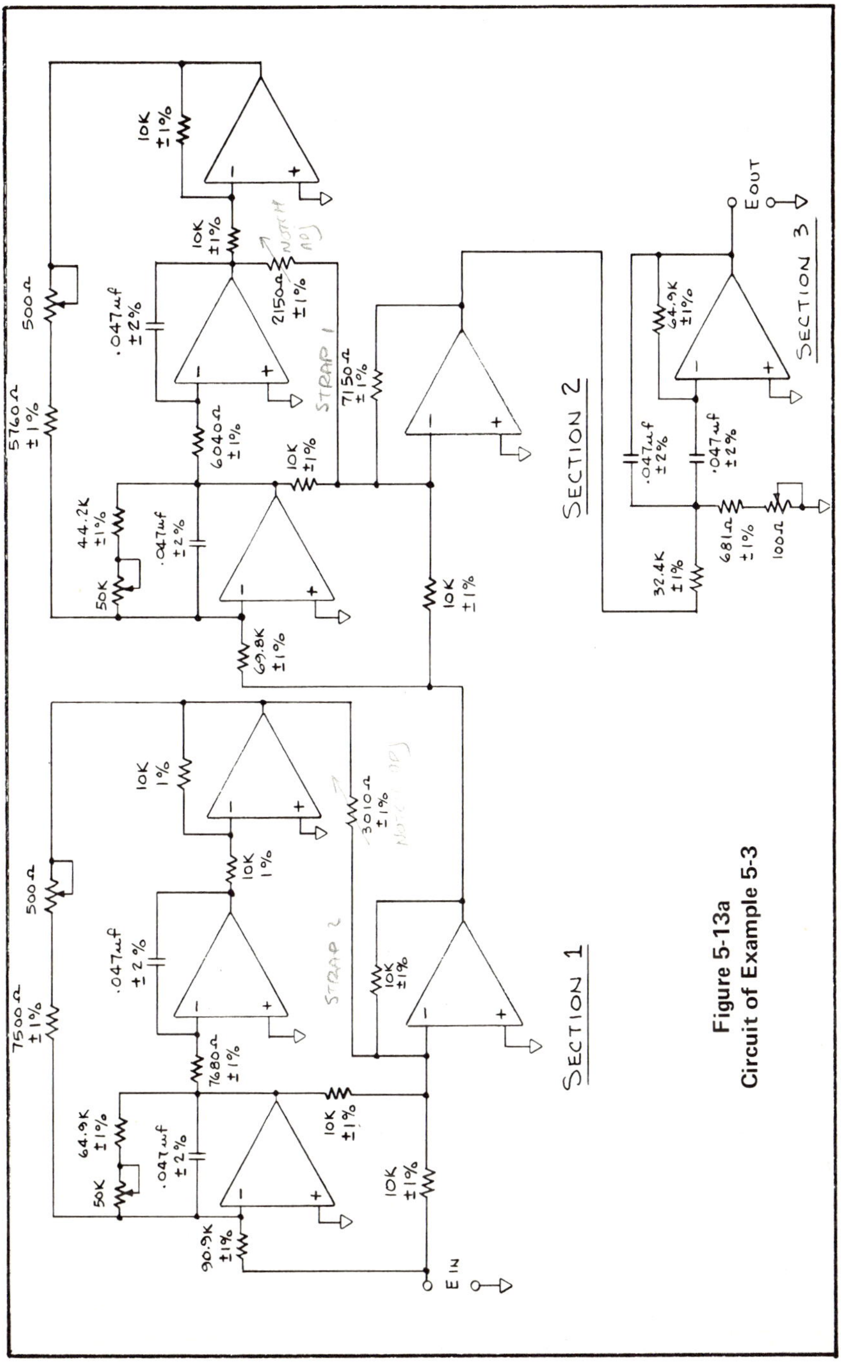

Figure 5-13a
Circuit of Example 5-3

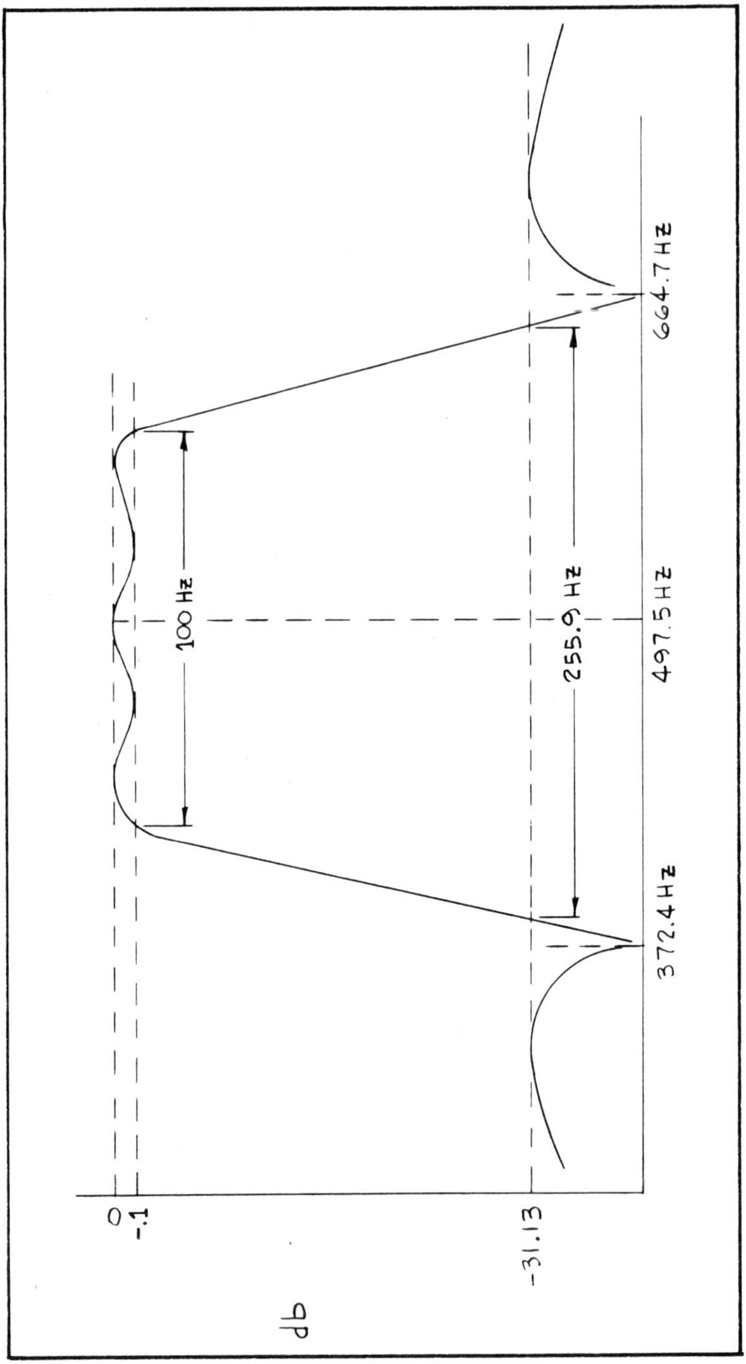

Figure 5-13b
Frequency Response of Example 5-3

well into the stop-band. If a delay line is required, a Low-Pass filter would not be the most efficient approach since the entire flat region couldn't be utilized.

The excellent delay properties of this family can be taken full advantage of with an All-Pass implementation of the Transfer Function. The polynomial roots of table 7.3d can be realized using the All-Pass section of Figure 5.6 for each complex pole-pair and the All-Pass section of figure 5-9 for each real pole.

Table 2.1 lists the delay at DC, T(o), the Radian frequencies, U, for 1% and 10% deviation from the zero frequency delay, and the T(o)U product.

The zero frequency delay is twice the value indicated in the curves of figure 2-17 because the All-Pass realization produces an equal number of zeros, in addition to the poles, which doubles the delay.

TABLE 2.1 Maximally Flat Delay Characteristics (All-Pass Realization)

N	T(o)	1% Deviation		10% Deviation	
		U	T(o)U	U	T(o)U
2	2.72	.412	1.121	.801	2.179
3	3.50	.691	2.419	1.109	3.882
4	4.26	.906	3.860	1.333	5.679
5	4.84	1.120	5.421	1.554	7.521
6	5.40	1.304	7.042	1.737	9.380
7	5.90	1.478	8.720	1.912	11.28
8	6.34	1.647	10.44	2.079	13.18
9	6.78	1.794	12.16	2.227	15.10

To determine the required N for a given delay Tgd, and a maximum frequency of operation Fgd, compute:

$$TU = 2\pi \; Fgd \; Tgd \tag{5-55}$$

Determine the lowest order, N, that has a T(o)U in excess of TU, using table 2.1. Use either the 1% column or 10% column depending upon the allowable error.

After selecting the required N calculate:

$$T_{ratio} = \frac{T(o)}{Tgd} \tag{5-56}$$

Each pole can now be denormalized by simply multiplying the values of table 7.3d, for the required N, by the T_{ratio}. Every complex pole having a real part α and an imaginary part β becomes:

$$\alpha' = \alpha \, T_{ratio} \tag{5-57}$$

$$\beta' = \beta \, T_{ratio} \tag{5-58}$$

A normalized real pole α becomes:

$$\alpha'' = \alpha \, T_{ratio} \tag{5-59}$$

To realize each complex pole, using the section of figure 5-6, first compute:

$$F_o = \frac{\sqrt{(\alpha')^2 + (\beta')^2}}{2\pi} \tag{5-60}$$

then

$$R_1 = \frac{\alpha'}{(2\pi F_o)^2 C} \tag{5-61}$$

$$R_2 = \frac{1}{\alpha' C} \tag{5-62}$$

$$R_3 = \frac{R\,R_2}{4\,R_1} \tag{5-63}$$

where C and R can be arbitrarily chosen.

R_1 can be made adjustable for very precise requirements. The tuning method of section 5.2b can then be used.

A denormalized real pole, which occurs for odd Ns′ only, can be satisfied by the circuit of figure 5-9 where:

$$R = \frac{1}{\alpha'' C} \tag{5-64}$$

R′ and C can be arbitrarily selected.

To adjust this section, R can be made variable. Monitor the section phase shift and adjust R until a phase shift of 90° is measured at a frequency of $1/(2\pi RC)$, in Hz. Either a phase meter or Lissajous Pattern may be used. The Lissajous Pattern is a circle or vertical or horizontal open ellipse for 90°.

Using 1% resistors and 2% capacitors, a delay accuracy of 3% is obtainable. This is sufficient for most requirements so adjustment will normally not be necessary.

Group Delay is generally measured using a Delay measuring instrument. However, where the delay changes gradually and smoothly, an alternate method can be employed. This method can be applied to All-Pass Delay Lines.

Record each frequency beginning at DC where the delay line has a phase shift that is a multiple of 180°. The Lissajous Pattern will display a closed Ellipse at each 180° multiple. Take each pair of adjacent frequencies, F_N and F_{N+1}, and compute:

$$T = \frac{1}{2\,(F_{N+1} - F_N)} \qquad (5\text{-}65)$$

This value is the nominal Group-Delay over each frequency interval. The less the frequency difference between 180° points, the more accurate this approximation becomes. For Low Order delay Lines ($N < 5$) this method has poor accuracy since too few 180° points are produced within the band of constant delay.

The following example demonstrates the design of an All-Pass Delay Line:

Example: 5-4

Given: Delay Line Requirement
Delay of 200 Microseconds ±4%
Maximum Operating Frequency of 1.5KHz
Flat Amplitude Response

Result: a. Determine Order N

Compute TU = 1.885 (5-55)

Table 2.1 indicates that for a delay deviation of 1%, N=3 has a T(o)U of 2.419 and a T(o) of 3.50.

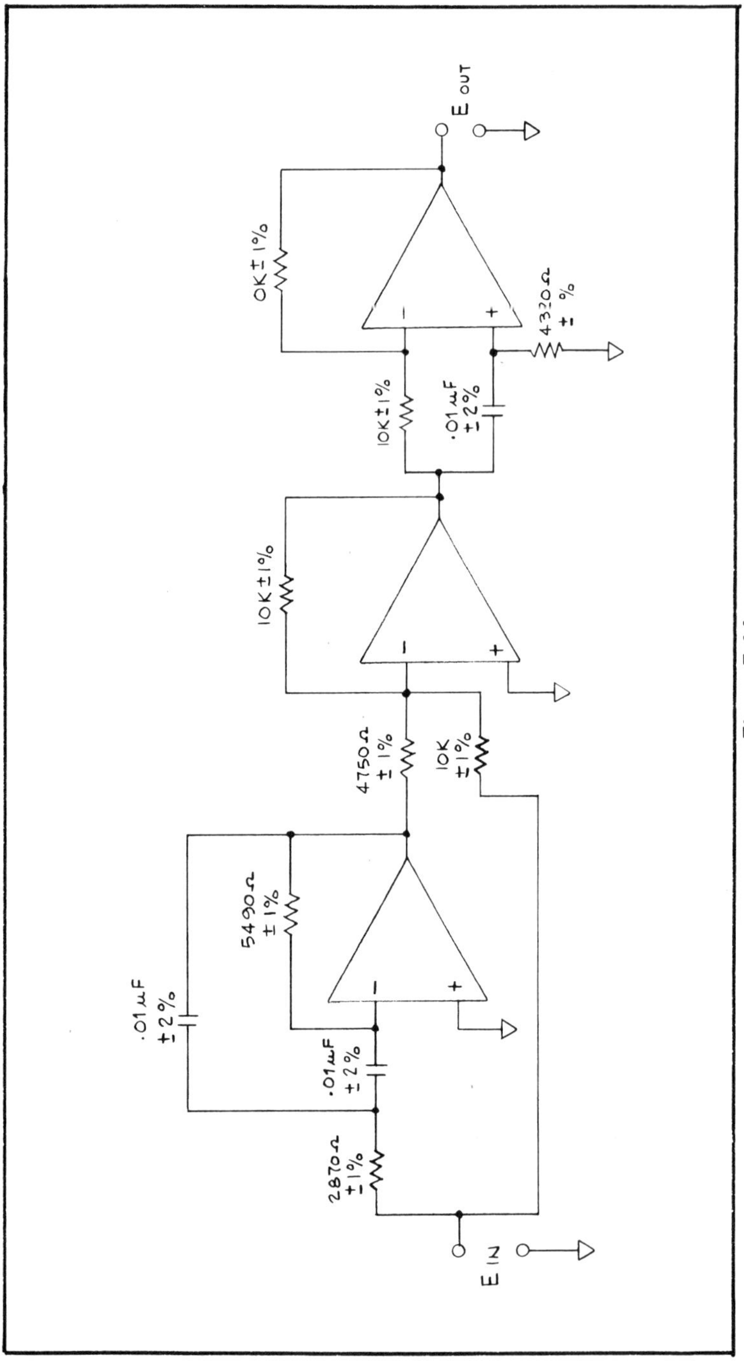

Figure 5-14a
Circuit of Example 5-4

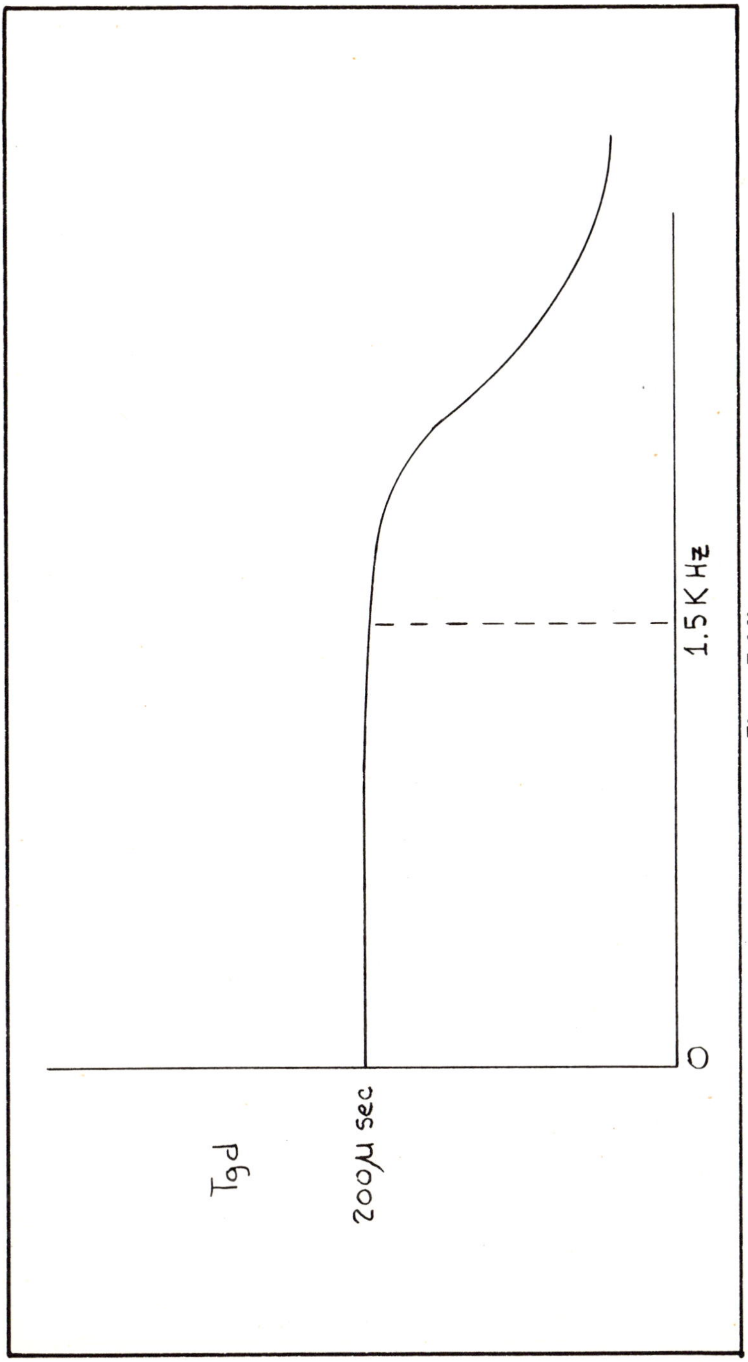

Figure 5-14b
Delay of Example 5-4

b. Denormalize Poles of Table 7.3d

Complex Pole-Pair $\alpha = 1.0509 \quad \beta = 1.0025$

Real Pole $\alpha = 1.3270$

Compute $T_{ratio} = 17{,}500$ (5-56)

Denormalized Complex Pole-Pair

$\alpha' = 18{,}391$ (5-57)

$\beta' = 17{,}544$ (5-58)

Denormalized Real Pole

$\alpha'' = 23{,}223$ (5-59)

c. Calculate Complex-Pole Section Values for circuit of Figure 5-6

(select C = .01uf, R = 10K)

$F_o = 4045\text{Hz}$ (5-60)
$R_1 = 2847\Omega$ (5-61)
$R_2 = 5437\Omega$ (5-62)
$R_3 = 4774\Omega$ (5-63)

d. Calculate Real-Pole Section Values for circuit of figure 5-9

(select C=.01uf, R'=10K)

$R = 4306\Omega$ (5-64)

Figure 5-14 illustrates the final circuit and response.

When the TU value computed by equation 5-55 is in excess of the highest T(o)U of table 2.1, the delay requirement can be subdivided into N smaller increments and realized by N delay Lines in cascade since delays add algebraically.

CHAPTER 6

Realizability and Practical Considerations

6.1 SENSITIVITY TO TOLERANCES

The concept of sensitivity can be broadly defined as the change in performance for a given change in a component value. Certain circuit configurations are extremely sensitive and may exhibit severe distortions of frequency response for small component changes. Figure 6-1 illustrates some possible extremes. Fortunately, high sensitivity circuits have been avoided in this book. Extremes such as those shown in Figure 6-1 will only occur with Large component errors or when operating beyond the capabilities of the operational amplifier.

Section 1.3 indicated that the cutoff of a Low-Pass filter can be frequency shifted inversely to all the resistor or all the capacitor values. A 1% resistor change will shift the filter cutoff by 1% or a 2% capacitor change will create a 2% change in cutoff. Let us then assume that all the resistors have a 1% tolerance and that all the capacitor tolerances are 2%. Let us further assume that all the components are simultaneously at the high or low extreme of their tolerance. The filter response will then be displaced by 3% inversely to the component change. Figure 6-2 shows the two extremes.

Although the assumption that all components will be at the high or low end of their tolerance simultaneously is not always valid, it serves to predict the possible boundaries of the response. Even if the components have a random distribution within their tolerance band the response will still remain within the general confines of these boundaries.

A 3% response error is acceptable for most Low-Pass or High-Pass filter requirements so 1% resistors and 2% capacitors are satisfactory.

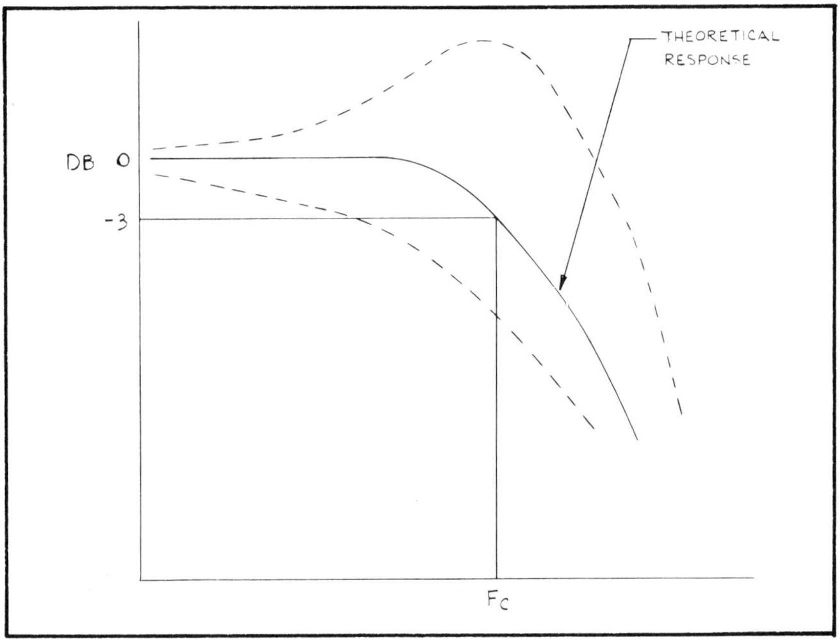

Figure 6-1
Possible Effects of Component Tolerances in High-Sensitivity Active Low-Pass Filter

For more critical requirements adjustable configurations are used so these tolerances are still acceptable.

Band-Pass filter requirements are much less tolerant of component errors. A 3% filter response shift may locate the desired pass-band on the slope of the frequency response if the filter is narrow. A random component distribution may severely distort the pass-band response shape as shown in Figure 4-5. Both of these effects are emphasized as the filter design becomes narrower. Since capacitor tolerances of less than 2% and resistor tolerances better than 1% severely increase component costs and since tuning errors due to amplifier shortcomings have to be contended with anyway, narrow filters should be made adjustable. Some guidelines are established in section 6.4.

6.2 AMPLIFIER LIMITATIONS

Amplifier shortcomings limit the useful range of active filter pass-bands to the audio frequency range. Once the required stop-band attenuation is established, the amplifier characteristics are of minor consequence beyond this range.

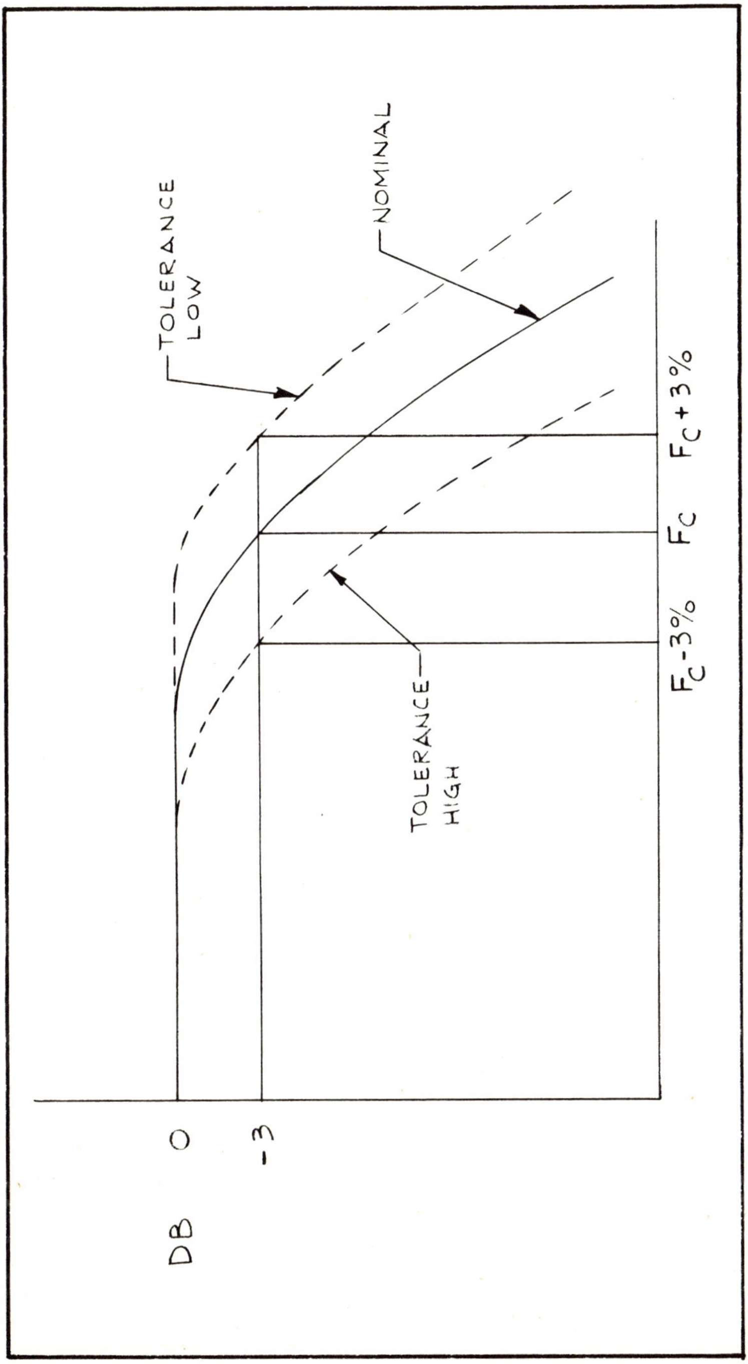

Figure 6-2
Effect of 1% Resistors and
2% Capacitors at Tolerance Extremes

The all-pole Low-Pass and High-Pass filters use operational amplifiers as unity-gain amplifiers (Voltage Followers). The various Elliptic Function and all-pole band-pass configurations use amplifiers as inverting or non-inverting gain elements. All of these filter configurations assume that the amplifier is ideal with the following three Open-Loop features:

A. Infinite Gain with no Phase-Shift

B. Infinite Input Impedance

C. Zero Output Impedance

As the amplifier characteristics deteriorate, the circuit poles and zeros are displaced, which results in a distortion of the frequency response shape.

a. Open-Loop Gain and Phase Shift

The most critical factor is the open-loop gain. In order to maintain amplifier stability the amplifier gain is band-limited by introducing a break-point in the frequency response so that the gain rolls off at approximately 6db per octave. The phase shift increases as the gain is reduced.

If we define A as the open Loop gain magnitude at the frequency of interest and β, the feedback factor, as the fraction of the output that is fed back to the input, the closed Loop gain, A_c, can be defined as:

$$A_c = \frac{A}{1 + A\beta} \qquad (6\text{-}1)$$

Equation 6-1 is the general equation for the gain of a negative-feedback system and should be familiar to those who have studied feedback systems or servo theory.

To determine the effect of open-Loop gain on the closed-Loop performance let us analyze the non-inverting amplifier circuit of figure 3-8. Substituting $R_1/(R_1 + R_2)$ for β in equation 6-1 and evaluating the effect of open Loop gain, A, for various design values of closed-Loop gain, A_c', figure 6-3 indicates the percent gain error. The $A_c' = 1$ condition corresponds to the Voltage Follower of figure 3-8a whereas the remaining gains are associated with figure 3-8b ($A_c' = 1 + R_2/R_1$).

It is clear that there is a diminishing return on closed-Loop gain accuracy as the open-Loop gain is increased. An open-Loop to closed-Loop gain ratio of 100 to 1 is more than adequate for normal filter requirements. If the feedback network is reactive, as in the circuit of

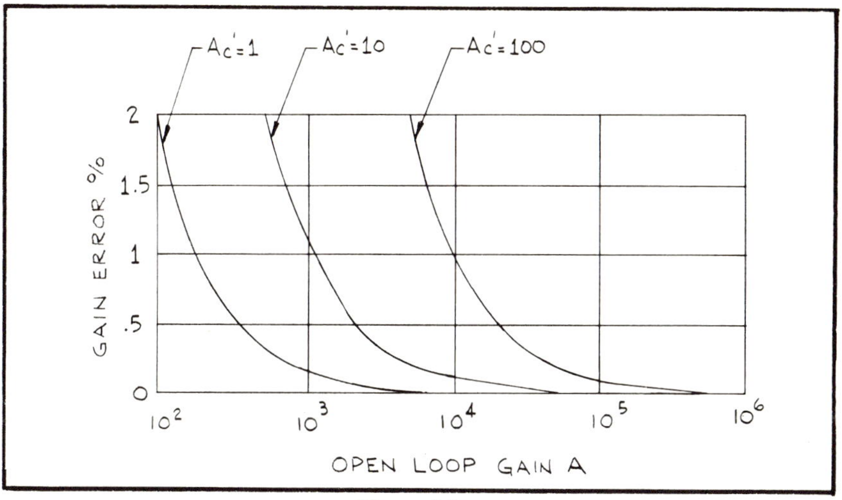

Figure 6-3
Effect of Open-Loop Gain on Gain Error

figure 4-1 or 4-2, the effect of open-Loop gain diminishes further so a gain ratio of 10 to 1 will provide acceptable accuracy. The open-Loop gain should then be at least $20Q^2$ for either figure 4-1 or 4-2. If the circuit is made adjustable, as in figure 4-6, a much lower open-Loop gain can be tolerated providing it is in excess of $2Q^2$.

The State-Variable circuits of figures' 4-3 and 5-10 are the least sensitive to amplifier open-loop gain of all configurations presented. Extremely high precision filters can be acheived with open-loop gains of 100 Q. Where adjustments are provided, open-loop gains as low as 10Q yield acceptable results.

The open-loop gain requirement of the single amplifier band-pass filter section of figures' 4-1 and 4-2 is proportional to Q^2 whereas the State-Variable circuit gain requirement is proportional to Q making this circuit more suitable for high-Q realizations.

As the open-loop gain decreases with frequency, the amplifier phase-shift increases. This creates an effect in the State-Variable circuit called Q Enhancement where the lagging phase shift will cause the circuit Q to increase. This increased Q can be approximated by:

$$Q' = \frac{Q}{1 + \frac{2Q\,(F_{3db} - 2F)}{A_o\,F_{3db}}} \qquad (6\text{-}2)$$

Where Q is the design Q, F_{3db} is the open-loop 3 db frequency, F is the frequency of interest generally the F_o' of the section, and A_o is the amplifier open-loop gain at zero frequency. If more than a 5% Q error is estimated by formula 6-2, a Q adjustment may be desired, as discussed in section 4.3b.

The closed-loop phase-shift of an amplifier can be estimated by:

$$\phi_c = \text{TAN}^{-1} \frac{\frac{F}{F_{3db}}}{1 + A\beta} \qquad (6\text{-}3)$$

For the single amplifier All-pole Low-Pass and High-Pass circuits of section 3.1, ϕ_c should be less than 4 degrees for the two-pole section and less than 6 degrees for the three-pole section. The β of the voltage follower is unity. F, the frequency of interest, should be the 3db cutoff of the filter, F_c.

The Elliptic-Function circuits of section 3.2 should have a ϕ_c of less than 4 degrees where F is the filter cutoff F_c, and β is 1/K where K is the tabulated section value from table 7.2.

The single-amplifier band-pass sections of figures 4-1 and 4-2 should have a ϕ_c not in excess of 4 degrees where F is F_o and β equals $1/(2Q^2)$. The adjustable version as in figure 4-6 can tolerate somewhat larger phase shifts.

b. Input and Output Impedance

Amplifier input and output impedance is usually not a problem in active filter realizations. The basic consideration is that the input impedance be much higher than the largest resistor connected to the amplifier input terminals and that the output impedance be much lower than the lowest resistor connected to the output terminal.

If the amplifier has an open-loop input impedance Z_i, the closed-loop value can be calculated by:

$$Z_{IN} = (1 + A\beta)\, Z_i \qquad (6\text{-}4)$$

Since the open-loop gain, A, is typically a few thousand, β is near unity and Z_i is around 50,000 ohms, the Z_{IN} computes to be in the range of 50 or 100 Megohms which is much higher than the normal impedance levels used in an active filter.

The output impedance is computed by:

$$Z_{OUT} = \frac{Z_o}{1 + A\beta} \qquad (6\text{-}5)$$

Since Z_o, the open-loop output impedance, is typically 200 ohms, Z_{OUT} is calculated to be around .1 ohms which of course is insignificant.

c. Dynamic Range

The output swing of an operational amplifier is limited to within a few volts of the supply voltages. Overdriving will result in clipping which prevents normal filter operation. The section gain or input level must be adjusted accordingly so as to restrict operation to the Linear region. It is good practice to observe the output of each amplifier on an oscilloscope while sweeping the filter input over the frequency range of interest at the maximum operating input level, to insure linear operation.

An additional problem is a possible detuning of Narrow Band Band-Pass Filters as the output level becomes Large. This effect occurs because of "Slew Rate Limiting" which is the inability of the amplifier to provide the steep output slope required for Large outputs at high frequencies.

The lower end of the dynamic range is limited by output noise. For maximum signal to noise ratio, Band-Pass filters should have maximum AC signal gain. This is especially important for the section nearest the input. A dynamic range of over 80db is obtainable.

d. Amplifier Selection

Although an extensive survey of operational amplifiers is outside the scope of this book, the two most popular types will be discussed. These are the familiar 709 and 741 monolithic operational amplifiers.

The 741 has a fixed amount of internal frequency compensation which insures stability for all closed-Loop gains. However, when the closed-Loop gain is greater than unity, less compensation is actually required. The fixed worst-case compensation produces a more severe open-Loop gain reduction than necessary at the high frequencies.

The 709 Operational Amplifier uses external frequency compensation which provides an additional degree of freedom. Less compensation can then be used for the higher gain configurations therefore increasing open-Loop bandwidth.

However, the 709 has a number of short-comings. The inverting input is highly susceptible to "Latchup" when connected as a Voltage Follower. During Latchup the input transistor saturates and converts the inverting input to non-inverting which locks up the output. This effect can be prevented by introducing a 10K resistor in the output to inverting-input feedback path instead of a direct connection.

The 709 has an input differential rating of 5 volts maximum and cannot tolerate an accidental output short circuit to ground for more than 5 seconds. The 741 can withstand up to a 30 volt input differential and will not be damaged by an indefinite short circuit.

Since the 709 compensation can be tailored for optimized high-frequency performance this amplifier is recommended for circuits requiring high open-Loop gain at frequencies where the 741 gain is inadequate. A typical 709 application might be for the single-amplifier band-pass filter circuit of section 4-1a for Qs' in excess of 10 at frequencies above a few KHz.

Unity or near unity amplifier gains can be readily acheived using the 741 operational amplifier at frequencies up to 10 or 20 KHz. Beyond this range the 709 will be slightly superior.

Above the audio range neither the 709 or the 741 amplifiers are adequate. Numerous other amplifiers are available which exhibit wider bandwidths at some additional cost. Probably the most popular type is the UA715 Wide-Band Operational Amplifier. This device will provide reasonable gain up to 500KHz at moderate cost. Alternate selections can be made by applying the established guidelines to the vendor data sheets.

Care should be taken with wide-band amplifiers, such as a lightly compensated UA709 or a UA715, to insure stability. Supply voltage connections should be adequately decoupled. Lead length should be minimized. Do not use less compensation than the manufacturers' recommendations.

The UA741 type will always remain stable and is highly recommended wherever possible.

Recently a new type of Low Cost amplifier has been introduced called an LM3900 developed mainly with the automotive industry in mind. It consists of a quad of amplifiers which operate in a manner which is somewhat unique when compared to the more conventional operational amplifiers. Whereas the conventional amplifiers amplify differential input voltages, this device amplifies differential input currents and can operate directly from a single supply voltage. It is internally compensated similar to a 741 and can be used for active filters with proper biasing.

Another recent development has been "Universal Active Filter Modules". These are basically "State Variable" building blocks similar to Figure 5-10 having multiple outputs and can be configured for a low-pass, high-pass, band-pass or band-reject response. External resistors are used to trim frequencies. Their major advantage over the discrete approach is size reduction. They offer no economic advantage.

The following table contains a tabulation of the more popular amplifier types and some of their significant characteristics. Depending on the manufacturer there may be slight variances in specifications and part numbers. Most devices are available in both T05 (round) packages and dual-in-line 14 pin packages. For each device there are normally two sets of specifications, Military (-55° C to +125° C) and commercial (0° C to +70° C). The tabulated parameters are for the commercial grade.

6.3 PASSIVE COMPONENT SELECTION

After a filter is designed and an amplifier has been selected, resistor and capacitor types must be chosen. Some of the major factors to be considered are Tolerance, Temperature coefficient, Temperature range, Retrace, and Physical size. Cost and availability also are important factors.

The effects of tolerances have been discussed in section 6.1 and need not be repeated. Component temperature coefficients may introduce serious tuning errors at temperature extremes even though resonances may have been adjusted at room temperature. Exceeding the rated temperature range of some components may permanently change or even destroy them.

a. Resistor Selection

Carbon composition resistors are unsuitable for active filters except where the requirements are extremely loose. They have a tolerance of 5 or 10% and are unstable with temperature. Also, the resistor values available are too far apart for most design requirements.

Precision Metal Film resistors are highly recommended for Active Filters and are readily available from all major electronic part distributors. They are stocked in a wide range of closely spaced values with a normal tolerance of 1%.

Device	LM324	709C	715	741	748	1458	LM3900
Input Offset Voltage (MV Max)	7	7.5	7.5	6	6	6	
Input Offset Current (NA Max)	50	500	250	200	200	200	
Input Bias Current (NA Max)	500	1500	1500	500	500	500	200
Voltage Gain at DC (Volts/V Min)	100k	25k	30k	20k	20k	20k	2.8k
Bandwidth for Voltage Follower(MHz)	1	1	10	1	1	1	2.5
Input Resistance		400k	1 Meg	1 Meg	1 Meg	1 Meg	1 Meg
Output Resistance		150	75	300	300	300	8k
Supply Voltage	±1.5 to ±15	±9 to ±18	±9 to ±18	±3 to ±18	±3 to ±18	±3 to ±18	±2 to ±18
Supply Current (MA Max)	2	6.6	5.5	2.9	5.6	2.9	10
Compensation	int.	ext.	ext.	int.	ext.	int.	int.
Comments	Quad, Single or Dual Supplies			Terminal for Offset Adjust	Terminal for Offset Adjust	Dual 741, No Offset Terminal	Quad Single or Dual Supplies

The military nomenclature is normally used to describe the part.

The part number is constructed as follows:

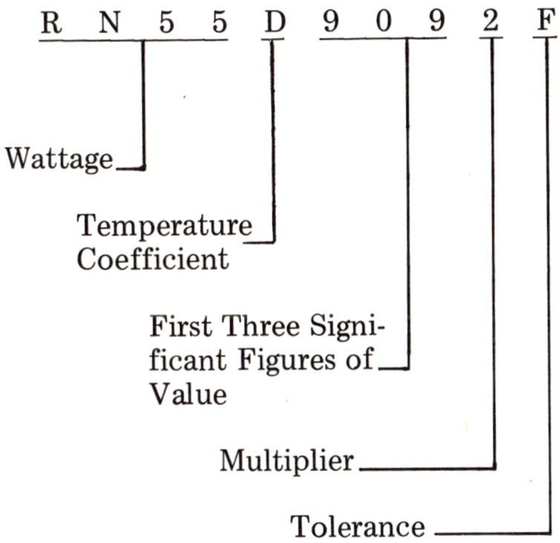

The wattage designator is either RN55 or RN60 representing ¼ WATT or ½ WATT respectively.

The temperature coefficient designator is either C for ±50 PPM/°C or D for $^{+200}_{-500}$ PPM/°C.

The value is described by its' three significant figures and the multiplier (number of Zeros that follow). The above example represents 90,900 ohms (90.9k).

The tolerance is either B for ±.1%, D for ±.5% or F for ±1%.

The RN55 type offers reduced size over the RN60 type and is comparable in cost and availability. Although the D temperature characteristic is somewhat poorer than the C characteristic, most resistors are manufactured to be far superior to their ratings so the D characteristic can be specified for most general requirements. The ±1% tolerance is adequate and is universally stocked whereas the tighter tolerance types usually must be ordered directly from the manufacturer.

The design values should be rounded off to the nearest standard value. The following table lists these standard values. The appropriate multiplier must be added.

TABLE 6.1
Standard Precision Resistor Values

100	162	261	422	681
102	165	267	432	698
105	169	274	442	715
107	174	280	453	732
110	178	287	464	750
113	182	294	475	768
115	187	301	487	787
118	191	309	499	806
121	196	316	511	825
124	200	324	523	845
127	205	332	536	866
130	210	340	549	887
133	215	348	562	909
137	221	357	576	931
140	226	365	590	953
143	232	374	604	976
147	237	383	619	
150	243	392	634	
154	249	402	649	
158	255	412	665	

b. Capacitor Selection

Selection of capacitor type is one of the more difficult decisions since many factors must be weighted against each other. Ideally, a capacitor type having a temperature coefficient equal and opposite to that of the resistors would result in a filter which would be independent of temperature. Since standard Metal-Film resistors may have either a positive or negative temperature coefficient attempts at matching the capacitors to the resistors will be fruitless. Resistors with rigidly controlled temperature coefficients can be obtained on special order from the manufacturer if absolutely required.

Polystyrene Capacitors

Polystyrene capacitors have excellent electrical properties which makes them highly suitable for active filters. They have a rigidly controlled temperature coefficient of -120 PPM/°C over the temperature range of -55°C to +85°C. Beyond +85°C, polystyrene capacitors will permanently and drastically change.

When a component is cycled over its temperature range it will never return exactly to its initial value when restored to room temperature. The amount of deviation due to temperature cycling is referred to as Retrace. Polystyrene Capacitors will typically retrace within .05%.

The practical range of values is from .0047uf to .1uf. Capacitors below .0047uf are extremely sensitive to direct heat from printed circuit soldering methods so they should be avoided to prevent production problems. Values above .1uf become extremely large and costly.

Capacitors exhibit Q or dissipation factor (1/Q) similar to inductors. Low Q capacitors will deteriorate circuit performance by introducing a shunt resistor of $Q/(2\pi FC)$. Polystyrene capacitors have Q's of over 2,000 so this effect can be neglected.

Voltage rating isn't critical since most active filters usually operate with small voltages in comparison to capacitor ratings. However, for stability with soldering, polystyrene capacitors less than .01uf should be rated at 200 Volts DC minimum. This will force a thicker dielectric film and therefore a capacitor less subject to permanent change due to excessive heat.

Polycarbonate Capacitors

In recent years, Polycarbonate capacitors have gained wide usage where precision capacitors are required. Since Polycarbonate film has a higher dielectric constant than Polystyrene, the practical capacity range can be extended to as high as 1uf.

The operating temperature ranges from -55° C to +125° C which is in excess of the +85° C limitation of Polystyrene. However the retrace is .25% which is somewhat inferior to Polystyrene. From +25° C to +55° C the temperature coefficient is typically zero within ±50 PPM/° C. Capacitor Q's usually exceed 500.

Mylar Capacitors

Mylar offers optimum space saving for large capacitance values. A Metalized Mylar capacitor is even smaller than the standard Mylar. Values as large as 5uf are then practical.

The operating temperature range is identical to Polycarbonate (-55° C to +125° C). The temperature coefficient is much higher normally averaging 400 PPM/° C. Capacitor Q's of 100 may deteriorate performance for high-Q filter circuits. Retrace is typically .3%.

Mica Capacitors

Mica capacitors are one of the most stable and reliable capacitor types available. They have a typical temperature coefficient of ±50 PPM/° C over the operating temperature range of -55° C to +150° C. Q's are in excess of 1000. Retrace is within .1%.

The most popular package is the Dipped Mica also known as the "DM" type. This construction features extremely small size and high reliability. The practical capacitance range is from 100pf to 5000pf. Values much below 100pf should be avoided since stray circuit capacities may introduce appreciable errors. Values in excess of 5000pf become costly.

Determining Capacitor Type and Value

For most applications Polystyrene capacitors are recommended. They offer excellent performance at reasonable cost.

For very low frequency filters, where capacitor values become Large, Polycarbonate should be used for highly critical requirements whereas the Mylar Type is recommended for general requirements.

Mica capacitors feature superior characteristics and minimum size. However their upper value limitation restricts their usage to above 1 or 2KHz.

TABLE 6.2
RETMA Standard Values

10	16	27	43	68
11	18	30	47	75
12	20	33	51	82
13	22	36	56	91
15	24	39	62	100

Precision capacitors are normally stocked by the manufacturer and can be ordered through a distributor. A vendor catalog will assist selection. Although special values can be obtained on special order it is desirable to restrict designs to standard values wherever possible. In designs where C can be arbitrarily chosen, standard values should be selected. The only consideration is that selection of too low a value may result in excessively high resistor values. As a guideline the largest resistor in a filter section should not exceed 500K ohms.

Standardization of capacitor values is in accordance with the RETMA (Radio Electronics-Television Manufacturers Association) series of two significant figures as listed in table 6.2. Tolerances of ±2% are adequate for most requirements.

When the capacitor values cannot be arbitrarily selected, such as in the Elliptic Function filters of section 3.2, non-standard values will result. These values can be constructed by paralleling standard values although it may prove more economical to order non-standard values. The spread of values (ratio of highest to lowest) cannot be controlled. However, the range of capacitance is determined by the chosen impedance level, Z, and can be readjusted as desired.

CHAPTER 7

Appendix

TABLE 7.1a

Butterworth Normalized Low-Pass Values

Order N	C1	C2	C3
2	1.414	0.7071	
3	3.546	1.392	0.2024
4	1.082 2.613	0.9241 0.3825	
5	1.753 3.235	1.354 0.3090	0.4214
6	1.035 1.414 3.863	0.9660 0.7071 0.2588	
7	1.531 1.604 4.493	1.336 0.6235 0.2225	0.4885
8	1.020 1.202 1.800 5.125	0.9809 0.8313 0.5557 0.1950	
9	1.455 1.305 2.000 5.758	1.327 0.7661 0.5000 0.1736	0.5170
10	1.012 1.122 1.414 2.202 6.390	0.9874 0.8908 0.7071 0.4540 0.1563	

TABLE 7.1b

.1db Ripple Chebyshev Normalized Low-Pass Values

Order N	C1	C2	C3
2	1.638	.6955	
3	6.653	1.825	.1345
4	1.900	1.241	
	4.592	.2410	
5	4.446	2.520	.3804
	6.810	.1580	
6	2.553	1.776	
	3.487	.4917	
	9.531	.1110	
7	5.175	3.322	.5693
	4.546	.3331	
	12.73	.08194	
8	3.270	2.323	
	3.857	.6890	
	5.773	.2398	
	16.44	.06292	
9	6.194	4.161	.7483
	4.678	.4655	
	7.170	.1812	
	20.64	.04980	
10	4.011	2.877	
	4.447	.8756	
	5.603	.3353	
	8.727	.1419	
	25.32	.04037	

TABLE 7.1c

.5db Ripple Chebyshev Normalized Low-Pass Values

Order N	C1	C2	C3
2	1.950	.6533	
3	11.23	2.250	.0895
4	2.582 6.233	1.300 .1802	
5	6.842 9.462	3.317 .1144	.3033
6	3.592 4.907 13.40	1.921 .3743 .07902	
7	7.973 6.446 18.07	4.483 .2429 .05778	.4700
8	4.665 5.502 8.237 23.45	2.547 .5303 .1714 .04409	
9	9.563 6.697 10.26 29.54	5.680 .3419 .1279 .03475	.6260
10	5.760 6.383 8.048 12.53 36.36	3.175 .6773 .2406 .09952 .02810	

TABLE 7.1d

Maximally Flat Delay Normalized Low-Pass Values

Order N	C1	C2	C3
2	.9066	.6800	
3	1.423	.9880	.2538
4	.7351	.6746	
	1.012	.3900	
5	1.010	.8712	.3095
	1.041	.3100	
6	.6352	.6100	
	.7225	.4835	
	1.073	.2561	
7	.8532	.7792	.3027
	.7250	.4151	
	1.100	.2164	
8	.5673	.5540	
	.6090	.4861	
	.7257	.3590	
	1.116	.1857	
9	.7564	.7070	.2851
	.6048	.4352	
	.7307	.3157	
	1.137	.1628	
10	.5172	.5092	
	.5412	.4682	
	.6000	.3896	
	.7326	.2792	
	1.151	.1437	

TABLE 7.2

ELLIPTIC FUNCTION NORMALIZED LOW-PASS VALUES

<u>Note:</u> All values are expressed in floating point E format. Conversion to fixed point decimal can be made as follows:

$$0.1000E+3 = 0.1000 \times 10^{3} = 100$$

$$0.1000E-2 = 0.1000 \times 10^{-2} = .001$$

N = 3	R_{db} = .01db	ω_s = 19.107	A_{db} = 74.93db
R1=0.3355E+00	C1=0.2503E+01	A=0.7784E+00	B=0.1607E+01
R2=0.6710E+00	C2=0.5562E+00	ZERO=0.2206E+02	
R3=0.3412E+02	C3=0.1093E-01		
R4=0.1536E+03	C4=0.5469E-02	K=0.1092E+01	GAIN=0.1085E+01
R5=0.1000E+01	C5=0.6381E+00	A=0.1567E+01	

N = 3	R_{db} = .01db	ω_s = 11.474	A_{db} = 61.61db
R1=0.3394E+00	C1=0.2478E+01	A=0.7726E+00	B=0.1608E+01
R2=0.6788E+00	C2=0.5506E+00	ZERO=0.1324E+02	
R3=0.1246E+02	C3=0.2997E-01		
R4=0.5611E+02	C4=0.1499E-01	K=0.1109E+01	GAIN=0.1089E+01
R5=0.1000E+01	C5=0.6354E+00	A=0.1574E+01	

N = 3	R_{db} = .01db	ω_s = 7.185	A_{db} = 49.33db
R1=0.3487E+00	C1=0.2419E+01	A=0.7584E+00	B=0.1609E+01
R2=0.6974E+00	C2=0.5375E+00	ZERO=0.8287E+01	
R3=0.5046E+01	C3=0.7428E-01		
R4=0.2271E+02	C4=0.3714E-01	K=0.1148E+01	GAIN=0.1098E+01
R5=0.1000E+01	C5=0.6289E+00	A=0.1590E+01	

N = 3	R_{db} = .01db	ω_s = 5.241	A_{db} = 41.00db
R1=0.3620E+00	C1=0.2340E+01	A=0.7378E+00	B=0.1610E+01
R2=0.7240E+00	C2=0.5199E+00	ZERO=0.6038E+01	
R3=0.2805E+01	C3-0.1342E+00		
R4=0.1262E+02	C4=0.6710E-01	K=0.1206E+01	GAIN=0.1111E+01
R5=0.1000E+01	C5=0.6193E+00	A=0.1615E+01	

N = 3	R_{db} = .01db	ω_s = 3.628	A_{db} = 31.14db
R1=0.3922E+00	C1=0.2183E+01	A=0.6898E+00	B=0.1610E+01
R2=0.7844E+00	C2=0.4851E+00	ZERO=0.4169E+01	
R3=0.1481E+01	C3=0.2570E+00		
R4=0.6662E+01	C4=0.1285E+00	K=0.1343E+01	GAIN=0.1141E+01
R5=0.1000E+01	C5=0.5968E+00	A=0.1676E+01	

N = 3	R_{db} = .01db	ω_s = 2.459	A_{db} = 20.4db
R1=0.4561E+00	C1=0.1930E+01	A=0.5809E+00	B=0.1602E+01
R2=0.9121E+00	C2=0.4290E+00	ZERO=0.2808E+01	
R3=0.8258E+00	C3=0.4738E+00		
R4=0.3716E+01	C4=0.2369E+00	K=0.1658E+01	GAIN=0.1212E+01
R5=0.1000E+01	C5=0.5438E+00	A=0.1839E+01	

N = 3	R_{db} = .1db	ω_s =11.47	A_{db}=71.24db
R1=0.3366E+00	C1=0.3425E+01	A=0.4832E+00	B=0.1208E+01
R2=0.6731E+00	C2=0.7610E+00	ZERO=0.1324E+02	
R3=0.2323E+02	C3=0.2205E-01		
R4=0.1045E+03	C4=0.1102E-01	K=0.1183E+01	GAIN=0.1171E+01
R5=0.1000E+01	C5=0.1024E+01	A=0.9757E+00	

N = 3	R_{db} = .1db	ω_s =8.206	A_{db}=62.45db
R1=0.3396E+00	C1=0.3395E+01	A=0.4804E+00	B=0.1208E+01
R2=0.6793E+00	C2=0.7545E+00	ZERO=0.9466E+01	
R3=0.1199E+02	C3=0.4274E-01		
R4=0.5395E+02	C4=0.2137E-01	K=0.1196E+01	GAIN=0.1174E+01
R5=0.1000E+01	C5=0.1021E+01	A=0.9792E+00	

N = 3	R_{db} = .1db	ω_s =5.241	A_{db}=50.63db
R1=0.3488E+00	C1=0.3310E+01	A=0.4720E+00	B=0.1210E+01
R2=0.6975E+00	C2=0.7356E+00	ZERO=0.6038E+01	
R3=0.5022E+01	C3=0.1021E+00		
R4=0.2260E+02	C4=0.5109E-01	K=0.1237E+01	GAIN=0.1182E+01
R5=0.1000E+01	C5=0.1010E+01	A=0.9899E+00	

N = 3	R_{db} = .1db	ω_s =3.628	A_{db}=40.77db
R1=0.3655E+00	C1=0.3166E+01	A=0.4564E+00	B=0.1213E+01
R2=0.7311E+00	C2=0.7036E+00	ZERO=0.4169E+01	
R3=0.2522E+01	C3=0.2040E+00		
R4=0.1134E+02	C4=0.1020E+00	K=0.1313E+01	GAIN=0.1197E+01
R5=0.1000E+01	C5=0.9897E+00	A=0.1010E+01	

N = 3	R_{db} = .1db	ω_s =2.559	A_{db}=31.13db
R1=0.3980E+00	C1=0.2924E+01	A=0.4249E+00	B=0.1216E+01
R2=0.7961E+00	C2=0.6497E+00	ZERO=0.2926E+01	
R3=0.1367E+01	C3=0.3783E+00		
R4=0.6152E+01	C4=0.1892E+00	K=0.1465E+01	GAIN=0.1227E+01
R5=0.1000E+01	C5=0.9484E+00	A=0.1054E+01	

N = 3	R_{db} = .1db	ω_s =1.788	A_{db}=20.53db
R1=0.4651E+00	C1=0.2540E+01	A=0.3539E+00	B=0.1219E+01
R2=0.9301E+00	C2=0.5645E+00	ZERO=0.2020E+01	
R3=0.7845E+00	C3=0.6692E+00		
R4=0.3530E+01	C4=0.3346E+00	K=0.1807E+01	GAIN=0.1295E+01
R5=0.1000E+01	C5=0.8544E+00	A=0.1170E+01	

N = 3	R_{db} = .18db	ω_s =11.47	A_{db}=73.81db
R1=0.3361E+00	C1=0.3698E+01	A=0.4182E+00	B=0.1131E+01
R2=0.6722E+00	C2=0.8219E+00	ZERO=0.1324E+02	
R3=0.2698E+02	C3=0.2047E-01		
R4=0.1214E+03	C4=0.1023E-01	K=0.1214E+01	GAIN=0.1204E+01
R5=0.1000E+01	C5=0.1185E+01	A=0.8434E+00	

N = 3	R_{db} = .18db	ω_s =7.185	A_{db}=61.54db
R1=0.3404E+00	C1=0.3653E+01	A=0.4149E+00	B=0.1132E+01
R2=0.6808E+00	C2=0.8117E+00	ZERO=0.8287E+01	
R3=0.1070E+02	C3=0.5161E-01		
R4=0.4818E+02	C4=0.2580E-01	K=0.1233E+01	GAIN=0.1208E+01
R5=0.1000E+01	C5=0.1179E+01	A=0.8478E+00	

N = 3	R_{db} = .18db	ω_s =4.810	A_{db}=50.92db
R1=0.3491E+00	C1=0.3564E+01	A=0.4081E+00	B=0.1134E+01
R2=0.6983E+00	C2=0.7920E+00	ZERO=0.5539E+01	
R3=0.4913E+01	C3=0.1125E+00		
R4=0.2211E+02	C4=0.5628E-01	K=0.1273E+01	GAIN=0.1215E+01
R5=0.1000E+01	C5=0.1167E+01	A=0.8567E+00	

N = 3	R_{db} = .18db	ω_s =3.236	A_{db}=40.23db
R1=0.3683E+00	C1=0.3383E+01	A=0.3927E+00	B=0.1137E+01
R2=0.7367E+00	C2=0.7519E+00	ZERO=0.3714E+01	
R3=0.2338E+01	C3=0.2369E+00		
R4=0.1052E+02	C4=0.1184E+00	K=0.1361E+01	GAIN=0.1231E+01
R5=0.1000E+01	C5=0.1139E+01	A=0.8776E+00	

N = 3	R_{db} = .18db	ω_s =2.281	A_{db}=30.41db
R1=0.4042E+00	C1=0.3095E+01	A=0.3627E+00	B=0.1142E+01
R2=0.8084E+00	C2=0.6877E+00	ZERO=0.2600E+01	
R3=0.1267E+01	C3=0.4387E+00		
R4=0.5703E+01	C4=0.2194E+00	K=0.1532E+01	GAIN=0.1263E+01
R5=0.1000E+01	C5=0.1085E+01	A=0.9213E+00	

N = 3	R_{db} = .18db	ω_s = 1.624	A_{db}=20.00db
R1=0.4743E+00	C1=0.2666E+01	A=0.2979E+00	B=0.1148E+01
R2=0.9485E+00	C2=0.5925E+00	ZERO=0.1824E+01	
R3=0.7479E+00	C3=0.7514E+00		
R4=0.3365E+01	C4=0.3757E+00	K=0.1895E+01	GAIN=0.1332E+01
R5=0.1000E+01	C5=0.9689E+00	A=0.1032E+01	

N = 3	R_{db} = .28db	ω_s =9.567	A_{db} = 71.10db
R1=0.3369E+00	C1=0.3901E+01	A=0.3693E+00	B=0.1079E+01
R2=0.6738E+00	C2=0.8670E+00	ZERO=0.1103E+02	
R3=0.2102E+02	C3=0.2780E-01		
R4=0.9457E+02	C4=0.1390E-01	K=0.1248E+01	GAIN=0.1235E+01
R5=0.1000E+01	C5=0.1340E+01	A=0.7465E+00	

N = 3	R_{db} = .28db	ω_s =6.393	A_{db} = 60.50db
R1=0.3413E+00	C1=0.3851E+01	A=0.3663E+00	B=0.1080E+01
R2=0.6826E+00	C2=0.8558E+00	ZERO=0.7370E+01	
R3=0.9492E+01	C3=0.6155E-01		
R4=0.4272E+02	C4=0.3077E-01	K=0.1268E+01	GAIN=0.1238E+01
R5=0.1000E+01	C5=0.1332E+01	A=0.7505E+00	

N = 3	R_{db} = .28db	ω_s =4.445	A_{db} = 50.86db
R1=0.3499E+00	C1=0.3758E+01	A=0.3604E+00	B=0.1082E+01
R2=0.6998E+00	C2=0.8350E+00	ZERO=0.5117E+01	
R3=0.4692E+01	C3=0.1245E+00		
R4=0.2111E+02	C4=0.6227E-01	K=0.1307E+01	GAIN=0.1245E+01
R5=0.1000E+01	C5=0.1318E+01	A=0.7585E+00	

N = 3	R_{db} = .28db	ω_s =2.924	A_{db} = 39.48db
R1=0.3719E+00	C1=0.3538E+01	A=0.3449E+00	B=0.1086E+01
R2=0.7438E+00	C2=0.7861E+00	ZERO=0.3350E+01	
R3=0.2142E+01	C3=0.2730E+00		
R4=0.9638E+01	C4=0.1365E+00	K=0.1410E+01	GAIN=0.1263E+01
R5=0.1000E+01	C5=0.1282E+01	A=0.7801E+00	

N = 3	R_{db} = .28db	ω_s = 2.130	A_{db} = 30.44db
R1=0.4069E+00	C1=0.3239E+01	A=0.3191E+00	B=0.1092E+01
R2=0.8137E+00	C2=0.7198E+00	ZERO=0.2423E+01	
R3=0.1229E+01	C3=0.4764E+00		
R4=0.5533E+01	C4=0.2382E+00	K=0.1578E+01	GAIN=0.1293E+01
R5=0.1000E+01	C5=0.1221E+01	A=0.8188E+00	

N = 3	R_{db} = .28db	ω_s = 1.556	A_{db} = 20.58db
R1=0.4739E+00	C1=0.2797E+01	A=0.2643E+00	B=0.1100E+01
R2=0.9479E+00	C2=0.6215E+00	ZERO=0.1742E+01	
R3=0.7490E+00	C3=0.7865E+00		
R4=0.3371E+01	C4=0.3933E+00	K=0.1927E+01	GAIN=0.1355E+01
R5=0.1000E+01	C5=0.1093E+01	A=0.9142E+00	

N = 5	Rdb = .01db	ω_s = 3.236	A_{db} = 72.10db
R1=0.3451E+00	C1=0.4265E+01	A=0.6469E+00	B=0.7873E+00
R2=0.6903E+00	C2=0.9478E+00	ZERO=0.5417E+01	
R3=0.6502E+01	C3=0.1006E+00		
R4=0.2926E+02	C4=0.5031E-01	K=0.8492E+00	GAIN=0.8202E+00
R1=0.3774E+00	C1=0.3222E+01	A=0.2286E+00	B=0.1212E+01
R2=0.7547E+00	C2=0.7160E+00	ZERO=0.3395E+01	
R3=0.1905E+01	C3=0.2837E+00		
R4=0.8572E+01	C4=0.1418E+00	K=0.1607E+01	GAIN=0.1420E+01
R5=0.1000E+01	C5=0.1189E+01	A=0.8405E+00	

N = 5	Rdb = .01db	ω_s = 2.559	A_{db} = 61.20db
R1=0.3531E+00	C1=0.4114E+01	A=0.6424E+00	B=0.8082E+00
R2=0.7062E+00	C2=0.9143E+00	ZERO=0.4240E+01	
R3=0.3970E+01	C3=0.1627E+00		
R4=0.1786E+02	C4=0.8133E-01	K=0.8866E+00	GAIN=0.8370E+00
R1=0.4030E+00	C1=0.3038E+01	A=0.2160E+00	B=0.1206E+01
R2=0.8059E+00	C2=0.6751E+00	ZERO=0.2681E+01	
R3=0.1286E+01	C3=0.4231E+00		
R4=0.5787E+01	C4=0.2116E+00	K=0.1731E+01	GAIN=0.1432E+01
R5=0.1000E+01	C5=0.1157E+01	A=0.8636E+00	

N = 5	R_{db} = .01db	ω_s = 2.000	A_{db} = 49.22db
R1=0.3686E+00	C1=0.3851E+01	A=0.6322E+00	B=0.8469E+00
R2=0.7371E+00	C2=0.8557E+00	ZERO=0.3251E+01	
R3=0.2325E+01	C3=0.2713E+00		
R4=0.1046E+02	C4=0.1357E+00	K=0.9609E+00	GAIN=0.8690E+00
R1=0.4452E+00	C1=0.2784E+01	A=0.1940E+00	B=0.1194E+01
R2=0.8904E+00	C2=0.6186E+00	ZERO=0.2089E+01	
R3=0.8843E+00	C3=0.6228E+00		
R4=0.3979E+01	C4=0.3114E+00	K=0.1941E+01	GAIN=0.1453E+01
R5=0.1000E+01	C5=0.1100E+01	A=0.9087E+00	

N = 5	R_{db} = .01db	ω_s = 1.701	A_{db} = 40.81db
R1=0.3866E+00	C1=0.3583E+01	A=0.6180E+00	B=0.8891E+00
R2=0.7732E+00	C2=0.7963E+00	ZERO=0.2709E+01	
R3=0.1613E+01	C3=0.3817E+00		
R4=0.7259E+01	C4=0.1908E+00	K=0.1050E+01	GAIN=0.9057E+00
R1=0.4848E+00	C1=0.2590E+01	A=0.1717E+00	B=0.1182E+01
R2=0.9695E+00	C2=0.5756E+00	ZERO=0.1772E+01	
R3=0.7114E+00	C3=0.7845E+00		
R4=0.3201E+01	C4=0.3923E+00	K=0.2145E+01	GAIN=0.1475E+01
R5=0.1000E+01	C5=0.1039E+01	A=0.9623E+00	

N = 5	R_{db} = .01db	ω_s = 1.414	A_{db} = 30.17db
R1=0.4239E+00	C1=0.3133E+01	A=0.5811E+00	B=0.9683E+00
R2=0.8479E+00	C2=0.6962E+00	ZERO=0.2166E+01	
R3=0.1039E+01	C3=0.5678E+00		
R4=0.4679E+01	C4=0.2839E+00	K=0.1247E+01	GAIN=0.9806E+00
R1=0.5443E+00	C1=0.2364E+01	A=0.1341E+00	B=0.1158E+01
R2=0.1088E+01	C2=0.5253E+00	ZERO=0.1465E+01	
R3=0.5733E+00	C3=0.9974E+00		
R4=0.2580E+01	C4=0.4987E+00	K=0.2471E+01	GAIN=0.1513E+01
R5=0.1000E+01	C5=0.9260E+00	A=0.1079E+01	

N = 5	R_{db} = .01db	ω_s = 1.2208	A_{db} = 20.04db
R1=0.4832E+00	C1=0.2617E+01	A=0.5039E+00	B=0.1073E+01
R2=0.9664E+00	C2=0.5815E+00	ZERO=0.1769E+01	
R3=0.7164E+00	C3=0.7845E+00		
R4=0.3224E+01	C4=0.3922E+00	K=0.1595E+01	GAIN=0.1100E+01
R1=0.6021E+00	C1=0.2209E+01	A=0.8944E-01	B=0.1124E+01
R2=0.1204E+01	C2=0.4909E+00	ZERO=0.1256E+01	
R3=0.4979E+00	C3=0.1187E+01		
R4=0.2240E+01	C4=0.5936E+00	K=0.2819E+01	GAIN=0.1561E+01
R5=0.1000E+01	C5=0.7689E+00	A=0.1301E+01	

N = 5	R_{db} = .1db	ω_s = 2.559	A_{db} = 70.83db
R1=0.3459E+00	C1=0.5271E+01	A=0.4336E+00	B=0.6991E+00
R2=0.6918E+00	C2=0.1171E+01	ZERO=0.4240E+01	
R3=0.6124E+01	C3=0.1323E+00		
R4=0.2756E+02	C4=0.6616E-01	K=0.1000E+01	GAIN=0.9639E+00
R1=0.3882E+00	C1=0.3553E+01	A=0.1505E+00	B=0.1077E+01
R2=0.7764E+00	C2=0.7896E+00	ZERO=0.2681E+01	
R3=0.1573E+01	C3=0.3898E+00		
R4=0.7077E+01	C4=0.1949E+00	K=0.1726E+01	GAIN=0.1482E+01
R5=0.1000E+01	C5=0.1755E+01	A=0.5698E+00	

N = 5	R_{db} = .1db	ω_s = 2.063	A_{db} = 60.39db
R1=0.3540E+00	C1=0.5060E+01	A=0.4307E+00	B=0.7181E+00
R2=0.7080E+00	C2=0.1124E+01	ZERO=0.3363E+01	
R3=0.3806E+01	C3=0.2092E+00		
R4=0.1713E+02	C4=0.1045E+00	K=0.1041E+01	GAIN=0.9808E+00
R1=0.4176E+00	C1=0.3315E+01	A=0.1410E+00	B=0.1074E+01
R2=0.8352E+00	C2=0.7366E+00	ZERO=0.2156E+01	
R3=0.1101E+01	C3=0.5585E+00		
R4=0.4956E+01	C4=0.2792E+00	K=0.1871E+01	GAIN=0.1493E+01
R5=0.1000E+01	C5=0.1698E+01	A=0.5887E+00	

N = 5	R_{db} = .1db	ω_s = 1.701	A_{db} = 50.44db
R1=0.3669E+00	C1=0.4757E+01	A=0.4252E+00	B=0.7468E+00
R2=0.7338E+00	C2=0.1057E+01	ZERO=0.2709E+01	
R3=0.2430E+01	C3=0.3192E+00		
R4=0.1093E+02	C4=0.1596E+00	K=0.1108E+01	GAIN=0.1006E+01
R1=0.4567E+00	C1=0.3046E+01	A=0.1274E+00	B=0.1070E+01
R2=0.9135E+00	C2=0.6768E+00	ZERO=0.1772E+01	
R3=0.8225E+00	C3=0.7517E+00		
R4=0.3701E+01	C4=0.3758E+00	K=0.2068E+01	GAIN=0.1509E+01
R5=0.1000E+01	C5=0.1616E+01	A=0.6186E+00	

N = 5	R_{db} = .1db	ω_s = 1.440	A_{db} = 40.90db
R1=0.3873E+00	C1=0.4346E+01	A=0.4144E+00	B=0.7889E+00
R2=0.7745E+00	C2=0.9659E+00	ZERO=0.2215E+01	
R3=0.1596E+01	C3=0.4688E+00		
R4=0.7181E+01	C4=0.2344E+00	K=0.1215E+01	GAIN=0.1046E+01
R1=0.5048E+00	C1=0.2776E+01	A=0.1090E+00	B=0.1064E+01
R2=0.1009E+01	C2=0.6170E+00	ZERO=0.1493E+01	
R3=0.6543E+00	C3=0.9518E+00		
R4=0.2945E+01	C4=0.4759E+00	K=0.2318E+01	GAIN=0.1531E+01
R5=0.1000E+01	C5=0.1501E+01	A=0.6662E+00	

N = 5	R_{db} = .1db	ω_s = 1.236	A_{db} = 30.59db
R1=0.4243E+00	C1=0.3756E+01	A=0.3891E+00	B=0.8570E+00
R2=0.8485E+00	C2=0.0347E+00	ZERO=0.1802E+01	
R3=0.1036E+01	C3=0.6831E+00		
R4=0.4666E+01	C4=0.3415E+00	K=0.1420E+01	GAIN=0.1115E+01
R1=0.5600E+00	C1=0.2531E+01	A=0.8245E-01	B=0.1054E+01
R2=0.1120E+01	C2=0.5625E+00	ZERO=0.1283E+01	
R3=0.5490E+00	C3=0.1147E+01		
R4=0.2470E+01	C4=0.5739E+00	K=0.2626E+01	GAIN=0.1563E+01
R5=0.1000E+01	C5=0.1323E+01	A=0.7558E+00	

N = 5	R_{db} = .1db	ω_s = 1.155	A_{db} = 24.92db
R1=0.4546E+00	C1=0.3380E+01	A=0.3635E+00	B=0.9060E+00
R2=0.9092E+00	C2=0.7511E+00	ZERO=0.1618E+01	
R3=0.8331E+00	C3=0.8197E+00		
R4=0.3749E+01	C4=0.4099E+00	K=0.1596E+01	GAIN=0.1170E+01
R1=0.5954E+00	C1=0.2401E+01	A=0.6525E-01	B=0.1047E+01
R2=0.1190E+01	C2=0.5335E+00	ZERO=0.1183E+01	
R3=0.5049E+00	C3=0.1258E+01		
R4=0.2272E+01	C4=0.6292E+00	K=0.2829E+01	GAIN=0.1584E+01
R5=0.1000E+01	C5=0.1196E+01	A=0.8355E+00	

N = 5	R_{db} = .18db	ω_s = 2.459	A_{db} = 71.50db
R1=0.3456E+00	C1=0.5564E+01	A=0.3812E+00	B=0.6805E+00
R2=0.6912E+00	C2=0.1236E+01	ZERO=0.4063E+01	
R3=0.6252E+01	C3=0.1367E+00		
R4=0.2813E+02	C4=0.6835E-01	K=0.1052E+01	GAIN=0.1015E+01
R1=0.3897E+00	C1=0.3635E+01	A=0.1320E+00	B=0.1050E+01
R2=0.7794E+00	C2=0.8078E+00	ZERO=0.2574E+01	
R3=0.1536E+01	C3=0.4099E+00		
R4=0.6912E+01	C4=0.2050E+00	K=0.1754E+01	GAIN=0.1500E+01
R5=0.1000E+01	C5=0.1994E+01	A=0.5016E+00	

N = 5	R_{db} = .18db	ω_s = 2.000	A_{db} = 61.43db
R1=0.3532E+00	C1=0.5347E+01	A=0.3789E+00	B=0.6980E+00
R2=0.7065E+00	C2=0.1188E+01	ZERO=0.3251E+01	
R3=0.3945E+01	C3=0.2128E+00		
R4=0.1775E+02	C4=0.1063E+00	K=0.1092E+01	GAIN=0.1030E+01
R1=0.4186E+00	C1=0.3392E+01	A=0.1238E+00	B=0.1049E+01
R2=0.8371E+00	C2=0.7538E+00	ZERO=0.2089E+01	
R3=0.1091E+01	C3=0.5782E+00		
R4=0.4911E+01	C4=0.2891E+00	K=0.1896E+01	GAIN=0.1510E+01
R5=0.1000E+01	C5=0.1930E+01	A=0.5180E+00	

N = 5	R_{db} = .18db	ω_s =1.624	A_{db} = 50.49db
R1=0.3673E+00	C1=0.4988E+01	A=0.3737E+00	B=0.7285E+00
R2=0.7345E+00	C2=0.1108E+01	ZERO=0.2566E+01	
R3=0.2405E+01	C3=0.3385E+00		
R4=0.1082E+02	C4=0.1693E+00	K=0.1165E+01	GAIN=0.1058E+01
R1=0.4626E+00	C1=0.3082E+01	A=0.1105E+00	B=0.1046E+01
R2=0.9251E+00	C2=0.6848E+00	ZERO=0.1690E+01	
R3=0.7954E+00	C3=0.7964E+00		
R4=0.3579E+01	C4=0.3982E+00	K=0.2119E+01	GAIN=0.1527E+01
R5=0.1000E+01	C5=0.1824E+01	A=0.5481E+00	

N = 5	R_{db} = .18db	ω_s =1.367	A_{db} = 40.23db
R1=0.3900E+00	C1=0.4500E+01	A=0.3629E+00	B=0.7738E+00
R2=0.7800E+00	C2=0.1000E+01	ZERO=0.2073E+01	
R3=0.1530E+01	C3=0.5098E+00		
R4=0.6885E+01	C4=0.2549E+00	K=0.1288E+01	GAIN=0.1100E+01
R1=0.5156E+00	C1=0.2780E+01	A=0.9226E-01	B=0.1042E+01
R2=0.1031E+01	C2=0.6177E+00	ZERO=0.1415E+01	
R3=0.6285E+00	C3=0.1013E+01		
R4=0.2828E+01	C4=0.5068E+00	K=0.2396E+01	GAIN=0.1549E+01
R5=0.1000E+01	C5=0.1676E+01	A=0.5968E+00	

$N = 5$	$R_{db} = .18db$	$\omega_s = 1.269$	$A_{db} = 35.13db$
R1=0.4067E+00	C1=0.4197E+01	A=0.3533E+00	B=0.8045E+00
R2=0.8135E+00	C2=0.9327E+00	ZERO=0.1872E+01	
R3=0.1231E+01	C3=0.6162E+00		
R4=0.5541E+01	C4=0.3081E+00	K=0.1380E+01	GAIN=0.1130E+01
R1=0.5448E+00	C1=0.2641E+01	A=0.8087E-01	B=0.1039E+01
R2=0.1089E+01	C2=0.5869E+00	ZERO=0.1309E+01	
R3=0.5725E+00	C3=0.1117E+01		
R4=0.2576E+01	C4=0.5586E+00	K=0.2555E+01	GAIN=0.1563E+01
R5=0.1000E+01	C5=0.1579E+01	A=0.6334E+00	

$N = 5$	$R_{db} = .18db$	$\omega_s = 1.192$	$A_{db} = 30.30db$
R1=0.4273E+00	C1=0.3876E+01	A=0.3398E+00	B=0.8395E+00
R2=0.8546E+00	C2=0.8613E+00	ZERO=0.1706E+01	
R3=0.1010E+01	C3=0.7285E+00		
R4=0.4547E+01	C4=0.3642E+00	K=0.1495E+01	GAIN=0.1166E+01
R1=0.5727E+00	C1=0.2523E+01	A=0.6876E-01	B=0.1035E+01
R2=0.1145E+01	C2=0.5607E+00	ZERO=0.1225E+01	
R3=0.5317E+00	C3=0.1207E+01		
R4=0.2393E+01	C4=0.6039E+00	K=0.2712E+01	GAIN=0.1578E+01
R5=0.1000E+01	C5=0.1471E+01	A=0.6796E+00	

N = 5	R_{db} = .28db	ω_s = 2.366	A_{db} = 71.71db
R1=0.3457E+00	C1=0.5785E+01	A=0.3410E+00	B=0.6681E+00
R2=0.6913E+00	C2=0.1286E+01		
		ZERO=0.3901E+01	
R3=0.6232E+01	C3=0.1426E+00		
R4=0.2804E+02	C4=0.7131E-01	K=0.1099E+01	GAIN=0.1060E+01
R1=0.3920E+00	C1=0.3683E+01	A=0.1176E+00	B=0.1032E+01
R2=0.7840E+00	C2=0.8185E+00		
		ZERO=0.2477E+01	
R3=0.1485E+01	C3=0.4321E+00		
R4=0.6683E+01	C4=0.2160E+00	K=0.1782E+01	GAIN=0.1516E+01
R5=0.1000E+01	C5=0.2224E+01	A=0.4497E+00	

N = 5	R_{db} = .28db	ω_s = 1.887	A_{db} = 60.51db
R1=0.3544E+00	C1=0.5519E+01	A=0.3386E+00	B=0.6879E+00
R2=0.7089E+00	C2=0.1226E+01		
		ZERO=0.3048E+01	
R3=0.3733E+01	C3=0.2329E+00		
R4=0.1680E+02	C4=0.1164E+00	K=0.1146E+01	GAIN=0.1077E+01
R1=0.4258E+00	C1=0.3397E+01	A=0.1091E+00	B=0.1031E+01
R2=0.8515E+00	C2=0.7548E+00		
		ZERO=0.1969E+01	
R3=0.1023E+01	C3=0.6280E+00		
R4=0.4606E+01	C4=0.3140E+00	K=0.1950E+01	GAIN=0.1526E+01
R5=0.1000E+01	C5=0.2140E+01	A=0.4673E+00	

N = 5	R_{db} = .28db	ω_s =1.556	A_{db} = 50.10db
R1=0.3685E+00	C1=0.5142E+01	A=0.3339E+00	B=0.7177E+00
R2=0.7370E+00	C2=0.1142E+01	ZERO=0.2438E+01	
R3=0.2330E+01	C3=0.3615E+00		
R4=0.1048E+02	C4=0.1807E+00	K=0.1220E+01	GAIN=0.1104E+01
R1=0.4698E+00	C1=0.3086E+01	A=0.9699E-01	B=0.1030E+01
R2=0.9396E+00	C2=0.6858E+00	ZERO=0.1617E+01	
R3=0.7651E+00	C3=0.8422E+00		
R4=0.3443E+01	C4=0.4211E+00	K=0.2173E+01	GAIN=0.1542E+01
R5=0.1000E+01	C5=0.2019E+01	A=0.4952E+00	

N = 5	R_{db} = .28db	ω_s =1.305	A_{db} = 39.17db
R1=0.3942E+00	C1=0.4572E+01	A=0.3226E+00	B=0.7673E+00
R2=0.7884E+00	C2=0.1015E+01	ZERO=0.1948E+01	
R3=0.1439E+01	C3=0.5564E+00		
R4=0.6477E+01	C4=0.2782E+00	K=0.1360E+01	GAIN=0.1149E+01
R1=0.5280E+00	C1=0.2757E+01	A=0.7860E-01	B=0.1027E+01
R2=0.1056E+01	C2=0.6127E+00	ZERO=0.1348E+01	
R3=0.6027E+00	C3=0.1073E+01		
R4=0.2712E+01	C4=0.5368E+00	K=0.2479E+01	GAIN=0.1565E+01
R5=0.1000E+01	C5=0.1831E+01	A=0.5462E+00	

N = 5	R_{db} = .28db	ω_s =1.236	A_{db} = 35.21db
R1=0.4078E+00	C1=0.4317E+01	A=0.3154E+00	B=0.7915E+00
R2=0.8157E+00	C2=0.9592E+00	ZERO=0.1802E+01	
R3=0.1216E+01	C3=0.6433E+00		
R4=0.5474E+01	C4=0.3216E+00	K=0.1435E+01	GAIN=0.1173E+01
R1=0.5509E+00	C1=0.2648E+01	A=0.7036E-01	B=0.1025E+01
R2=0.1101E+01	C2=0.5884E+00	ZERO=0.1273E+01	
R3=0.5627E+00	C3=0.1152E+01		
R4=0.2532E+01	C4=0.5761E+00	K=0.2604E+01	GAIN=0.1575E+01
R5=0.1000E+01	C5=0.1743E+01	A=0.5737E+00	

N = 5	R_{db} = .28db	ω_s =1.166	A_{db} = 30.46db
R1=0.4284E+00	C1=0.3982E+01	A=0.3031E+00	B=0.8255E+00
R2=0.8568E+00	C2=0.8848E+00	ZERO=0.1647E+01	
R3=0.1001E+01	C3=0.7570E+00		
R4=0.4507E+01	C4=0.3785E+00	K=0.1551E+01	GAIN=0.1207E+01
R1=0.5781E+00	C1=0.2531E+01	A=0.5948E-01	B=0.1023E+01
R2=0.1156E+01	C2=0.5624E+00	ZERO=0.1196E+01	
R3=0.5249E+00	C3=0.1238E+01		
R4=0.2362E+01	C4=0.6194E+00	K=0.2756E+01	GAIN=0.1589E+01
R5=0.1000E+01	C5=0.1623E+01	A=0.6163E+00	

N = 7	R_{db} = .01db	ω_s = 1.662	A_{db} = 70.56db
R1=0.3499E+00	C1=0.5485E+01	A=0.5212E+00	B=0.5825E+00
R2=0.6998E+00	C2=0.1218E+01	ZERO=0.3508E+01	
R3=0.4699E+01	C3=0.1815E+00		
R4=0.2114E+02	C4=0.9076E-01	K=0.8162E+00	GAIN=0.7776E+00
R1=0.4114E+00	C1=0.3692E+01	A=0.2938E+00	B=0.9427E+00
R2=0.8229E+00	C2=0.8205E+00	ZERO=0.2040E+01	
R3=0.1170E+01	C3=0.5767E+00		
R4=0.5268E+01	C4=0.2883E+00	K=0.1568E+01	GAIN=0.1270E+01
R1=0.4722E+00	C1=0.2903E+01	A=0.8994E-01	B=0.1090E+01
R2=0.9443E+00	C2=0.6451E+00	ZERO=0.1696E+01	
R3=0.7558E+00	C3=0.8060E+00		
R4=0.3401E+01	C4=0.4030E+00	K=0.2206E+01	GAIN=0.1557E+01
R5=0.1000E+01	C5=0.1568E+01	A=0.6378E+00	

N = 7	R_{db} = .01db	ω_s = 1.466	A_{db} = 60.55db
R1=0.3579E+00	C1=0.5155E+01	A=0.5279E+00	B=0.6183E+00
R2=0.7157E+00	C2=0.1145E+01	ZERO=0.2997E+01	
R3=0.3242E+01	C3=0.2529E+00		
R4=0.1459E+02	C4=0.1265E+00	K=0.8599E+00	GAIN=0.8009E+00
R1=0.4387E+00	C1=0.3429E+01	A=0.2740E+00	B=0.9588E+00
R2=0.8773E+00	C2=0.7620E+00	ZERO=0.1774E+01	
R3=0.9255E+00	C3=0.7223E+00		
R4=0.4165E+01	C4=0.3612E+00	K=0.1711E+01	GAIN=0.1300E+01
R1=0.5094E+00	C1=0.2713E+01	A=0.7932E-01	B=0.1082E+01
R2=0.1018E+01	C2=0.6028E+00	ZERO=0.1494E+01	
R3=0.6430E+00	C3=0.9551E+00		
R4=0.2893E+01	C4=0.4776E+00	K=0.2398E+01	GAIN=0.1569E+01
R5=0.1000E+01	C5=0.1482E+01	A=0.6746E+00	

N = 7	R_{db} = .01db	ω_s = 1.305	A_{db} = 50.11db
R1=0.3712E+00	C1=0.4706E+01	A=0.5345E+00	B=0.6722E+00
R2=0.7423E+00	C2=0.1045E+01	ZERO=0.2549E+01	
R3=0.2181E+01	C3=0.3560E+00		
R4=0.9813E+01	C4=0.1780E+00	K=0.9318E+00	GAIN=0.8369E+00
R1=0.4751E+00	C1=0.3128E+01	A=0.2456E+00	B=0.9792E+00
R2=0.9501E+00	C2=0.6950E+00	ZERO=0.1548E+01	
R3=0.7449E+00	C3=0.8865E+00		
R4=0.3352E+01	C4=0.4433E+00	K=0.1913E+01	GAIN=0.1342E+01
R1=0.5517E+00	C1=0.2532E+01	A=0.6569E-01	B=0.1071E+01
R2=0.1103E+01	C2=0.5626E+00	ZERO=0.1327E+01	
R3=0.5614E+00	C3=0.1105E+01		
R4=0.2526E+01	C4=0.5529E+00	K=0.2624E+01	GAIN=0.1585E+01
R5=0.1000E+01	C5=0.1366E+01	A=0.7322E+00	

N = 7	R_{db} = .01db	ω_s = 1.252	A_{db} = 45.93db
R1=0.3787E+00	C1=0.4492E+01	A=0.5361E+00	B=0.7002E+00
R2=0.7573E+00	C2=0.9982E+00	ZERO=0.2391E+01	
R3=0.1856E+01	C3=0.4072E+00		
R4=0.8353E+01	C4=0.2036E+00	K=0.9725E+00	GAIN=0.8561E+00
R1=0.4920E+00	C1=0.3004E+01	A=0.2314E+00	B=0.9881E+00
R2=0.9840E+00	C2=0.6675E+00	ZERO=0.1471E+01	
R3=0.6890E+00	C3=0.9533E+00		
R4=0.3101E+01	C4=0.4767E+00	K=0.2011E+01	GAIN=0.1363E+01
R1=0.5688E+00	C1=0.2468E+01	A=0.5954E-01	B=0.1066E+01
R2=0.1137E+01	C2=0.5485E+00	ZERO=0.1271E+01	
R3=0.5368E+00	C3=0.1162E+01		
R4=0.2415E+01	C4=0.5812E+00	K=0.2717E+01	GAIN=0.1592E+01
R5=0.1000E+01	C5=0.1310E+01	A=0.7634E+00	

N = 7	R_{db} = .01db	ω_s = 1.192	A_{db} = 40.54db
R1=0.3909E+00	C1=0.4187E+01	A=0.5363E+00	B=0.7429E+00
R2=0.7819E+00	C2=0.9305E+00	ZERO=0.2204E+01	
R3=0.1508E+01	C3=0.4825E+00		
R4=0.6786E+01	C4=0.2412E+00	K=0.1039E+01	GAIN=0.8862E+00
R1=0.5157E+00	C1=0.2846E+01	A=0.2106E+00	B=0.1000E+01
R2=0.1031E+01	C2=0.6325E+00	ZERO=0.1382E+01	
R3=0.6285E+00	C3=0.1038E+01		
R4=0.2828E+01	C4=0.5190E+00	K=0.2153E+01	GAIN=0.1392E+01
R1=0.5902E+00	C1=0.2396E+01	A=0.5116E-01	B=0.1059E+01
R2=0.1180E+01	C2=0.5325E+00	ZERO=0.1208E+01	
R3=0.5107E+00	C3=0.1230E+01		
R4=0.2298E+01	C4=0.6153E+00	K=0.2837E+01	GAIN=0.1602E+01
R5=0.1000E+01	C5=0.1229E+01	A=0.8130E+00	

N = 7	R_{db} = .01db	ω_s = 1.155	A_{db} = 36.61db
R1=0.4023E+00	C1=0.3946E+01	A=0.5342E+00	B=0.7795E+00
R2=0.8046E+00	C2=0.8768E+00	ZERO=0.2078E+01	
R3=0.1296E+01	C3=0.5442E+00		
R4=0.5834E+01	C4=0.2721E+00	K=0.1101E+01	GAIN=0.9129E+00
R1=0.5340E+00	C1=0.2736E+01	A=0.1934E+00	B=0.1008E+01
R2=0.1067E+01	C2=0.6079E+00	ZERO=0.1323E+01	
R3=0.5914E+00	C3=0.1097E+01		
R4=0.2661E+01	C4=0.5489E+00	K=0.2267E+01	GAIN=0.1415E+01
R1=0.6048E+00	C1=0.2351E+01	A=0.4482E-01	B=0.1053E+01
R2=0.1209E+01	C2=0.5225E+00	ZERO=0.1168E+01	
R3=0.4951E+00	C3=0.1276E+01		
R4=0.2228E+01	C4=0.6382E+00	K=0.2921E+01	GAIN=0.1610E+01
R5=0.1000E+01	C5=0.1165E+01	A=0.8581E+00	

N = 7	R_{db} = .1dB	ω_s = 1.390	A_{db} = 65.57dB
R1=0.3526E+00	C1=0.6345E+01	A=0.3640E+00	B=0.5631E+00
R2=0.7052E+00	C2=0.1410E+01	ZERO=0.2789E+01	
R3=0.4069E+01	C3=0.2444E+00		
R4=0.1831E+02	C4=0.1221E+00	K=0.9973E+00	GAIN=0.9429E+00
R1=0.4343E+00	C1=0.3761E+01	A=0.1899E+00	B=0.8984E+00
R2=0.8686E+00	C2=0.8358E+00	ZERO=0.1668E+01	
R3=0.9557E+00	C3=0.7596E+00		
R4=0.4301E+01	C4=0.3798E+00	K=0.1812E+01	GAIN=0.1391E+01
R1=0.5108E+00	C1=0.2845E+01	A=0.5480E-01	B=0.1030E+01
R2=0.1021E+01	C2=0.6322E+00	ZERO=0.1415E+01	
R3=0.6397E+00	C3=0.1009E+01		
R4=0.2879E+01	C4=0.5048E+00	K=0.2445E+01	GAIN=0.1596E+01
R5=0.1000E+01	C5=0.2164E+01	A=0.4620E+00	

N = 7	R_{db} = .1db	ω_s = 1.325	A_{db} = 61.17db
R1=0.3564E+00	C1=0.6141E+01	A=0.3665E+00	B=0.5791E+00
R2=0.7128E+00	C2=0.1365E+01	ZERO=0.2606E+01	
R3=0.3435E+01	C3=0.2832E+00		
R4=0.1546E+02	C4=0.1416E+00	K=0.1019E+01	GAIN=0.9536E+00
R1=0.4481E+00	C1=0.3619E+01	A=0.1827E+00	B=0.9066E+00
R2=0.8962E+00	C2=0.8043E+00	ZERO=0.1576E+01	
R3=0.8677E+00	C3=0.8307E+00		
R4=0.3905E+01	C4=0.4153E+00	K=0.1886E+01	GAIN=0.1403E+01
R1=0.5284E+00	C1=0.2755E+01	A=0.5107E-01	B=0.1029E+01
R2=0.1056E+01	C2=0.6122E+00	ZERO=0.1347E+01	
R3=0.6020E+00	C3=0.1074E+01		
R4=0.2709E+01	C4=0.5374E+00	K=0.2537E+01	GAIN=0.1601E+01
R5=0.1000E+01	C5=0.2099E+01	A=0.4764E+00	

N = 7	R_{db} = .1db	ω_s = 1.252	A_{db} = 55.56db
R1=0.3625E+00	C1=0.5846E+01	A=0.3697E+00	B=0.6035E+00
R2=0.7251E+00	C2=0.1299E+01	ZERO=0.2391E+01	
R3=0.2759E+01	C3=0.3414E+00		
R4=0.1241E+02	C4=0.1707E+00	K=0.1055E+01	GAIN=0.9701E+00
R1=0.4678E+00	C1=0.3432E+01	A=0.1721E+00	B=0.9183E+00
R2=0.9356E+00	C2=0.7626E+00	ZERO=0.1471E+01	
R3=0.7730E+00	C3=0.9230E+00		
R4=0.3479E+01	C4=0.4615E+00	K=0.1994E+01	GAIN=0.1421E+01
R1=0.5513E+00	C1=0.2647E+01	A=0.4585E-01	B=0.1026E+01
R2=0.1102E+01	C2=0.5882E+00	ZERO=0.1271E+01	
R3=0.5621E+00	C3=0.1153E+01		
R4=0.2529E+01	C4=0.5770E+00	K=0.2658E+01	GAIN=0.1607E+01
R5=0.1000E+01	C5=0.2006E+01	A=0.4985E+00	

N = 7	R_{db} = .1db	ω_s = 1.192	A_{db} = 50.17db
R1=0.3702E+00	C1=0.5524E+01	A=0.3724E+00	B=0.6318E+00
R2=0.7405E+00	C2=0.1227E+01	ZERO=0.2204E+01	
R3=0.2229E+01	C3=0.4078E+00		
R4=0.1003E+02	C4=0.2039E+00	K=0.1099E+01	GAIN=0.9897E+00
R1=0.4890E+00	C1=0.3248E+01	A=0.1600E+00	B=0.9306E+00
R2=0.9780E+00	C2=0.7219E+00	ZERO=0.1382E+01	
R3=0.6980E+00	C3=0.1011E+01		
R4=0.3141E+01	C4=0.5057E+00	K=0.2114E+01	GAIN=0.1441E+01
R1=0.5732E+00	C1=0.2553E+01	A=0.4038E-01	B=0.1024E+01
R2=0.1146E+01	C2=0.5673E+00	ZERO=0.1208E+01	
R3=0.5311E+00	C3=0.1224E+01		
R4=0.2390E+01	C4=0.6123E+00	K=0.2776E+01	GAIN=0.1614E+01
R5=0.1000E+01	C5=0.1904E+01	A=0.5251E+00	

N = 7	R_{db} = .1db	ω_s = 1.155	A_{db} = 46.24db
R1=0.3774E+00	C1=0.5264E+01	A=0.3737E+00	B=0.6561E+00
R2=0.7547E+00	C2=0.1169E+01	ZERO=0.2078E+01	
R3=0.1905E+01	C3=0.4635E+00		
R4=0.8571E+01	C4=0.2318E+00	K=0.1139E+01	GAIN=0.1006E+01
R1=0.5058E+00	C1=0.3115E+01	A=0.1499E+00	B=0.9402E+00
R2=0.1011E+01	C2=0.6922E+00	ZERO=0.1323E+01	
R3=0.6517E+00	C3=0.1074E+01		
R4=0.2933E+01	C4=0.5372E+00	K=0.2211E+01	GAIN=0.1457E+01
R1=0.5886E+00	C1=0.2491E+01	A=0.3613E-01	B=0.1022E+01
R2=0.1177E+01	C2=0.5536E+00	ZERO=0.1168E+01	
R3=0.5124E+00	C3=0.1272E+01		
R4=0.2306E+01	C4=0.6359E+00	K=0.2860E+01	GAIN=0.1620E+01
R5=0.1000E+01	C5=0.1823E+01	A=0.5486E+00	

N = 7	R_{db} = .18db	ω_s = 1.440	A_{db} = 71.20db
R1=0.3487E+00	C1=0.6859E+01	A=0.3197E+00	B=0.5396E+00
R2=0.6973E+00	C2=0.1524E+01	ZERO=0.2925E+01	
R3=0.5055E+01	C3=0.2103E+00		
R4=0.2275E+02	C4=0.1051E+00	K=0.1032E+01	GAIN=0.9870E+00
R1=0.4220E+00	C1=0.3966E+01	A=0.1731E+00	B=0.8792E+00
R2=0.8441E+00	C2=0.8814E+00	ZERO=0.1737E+01	
R3=0.1057E+01	C3=0.7036E+00		
R4=0.4758E+01	C4=0.3518E+00	K=0.1784E+01	GAIN=0.1409E+01
R1=0.4955E+00	C1=0.2960E+01	A=0.5122E-01	B=0.1021E+01
R2=0.9910E+00	C2=0.6578E+00	ZERO=0.1466E+01	
R3=0.6790E+00	C3=0.9602E+00		
R4=0.3055E+01	C4=0.4801E+00	K=0.2378E+01	GAIN=0.1600E+01
R5=0.1000E+01	C5=0.2512E+01	A=0.3980E+00	

N = 7	R_{db} = .18db	ω_s = 1.346	A_{db} = 65.20db
R1=0.3528E+00	C1=0.6593E+01	A=0.3229E+00	B=0.5581E+00
R2=0.7057E+00	C2=0.1465E+01	ZERO=0.2665E+01	
R3=0.4018E+01	C3=0.2573E+00		
R4=0.1808E+02	C4=0.1287E+00	K=0.1057E+01	GAIN=0.9989E+00
R1=0.4392E+00	C1=0.3774E+01	A=0.1654E+00	B=0.8897E+00
R2=0.8784E+00	C2=0.8386E+00	ZERO=0.1606E+01	
R3=0.9218E+00	C3=0.7992E+00		
R4=0.4148E+01	C4=0.3996E+00	K=0.1875E+01	GAIN=0.1423E+01
R1=0.5189E+00	C1=0.2831E+01	A=0.4703E-01	B=0.1020E+01
R2=0.1037E+01	C2=0.6291E+00	ZERO=0.1368E+01	
R3=0.6214E+00	C3=0.1050E+01		
R4=0.2796E+01	C4=0.5253E+00	K=0.2499E+01	GAIN=0.1605E+01
R5=0.1000E+01	C5=0.2420E+01	A=0.4132E+00	

N = 7	R_{db} = .18db	ω_s = 1.287	A_{db} = 60.91db
R1=0.3566E+00	C1=0.6376E+01	A=0.3252E+00	B=0.5739E+00
R2=0.7133E+00	C2=0.1417E+01	ZERO=0.2495E+01	
R3=0.3401E+01	C3=0.2972E+00		
R4=0.1530E+02	C4=0.1486E+00	K=0.1079E+01	GAIN=0.1009E+01
R1=0.4531E+00	C1=0.3629E+01	A=0.1589E+00	B=0.8981E+00
R2=0.9063E+00	C2=0.8065E+00	ZERO=0.1521E+01	
R3=0.8405E+00	C3=0.8696E+00		
R4=0.3782E+01	C4=0.4348E+00	K=0.1950E+01	GAIN=0.1434E+01
R1=0.5362E+00	C1=0.2743E+01	A=0.4371E-01	B=0.1018E+01
R2=0.1072E+01	C2=0.6095E+00	ZERO=0.1307E+01	
R3=0.5873E+00	C3=0.1113E+01		
R4=0.2643E+01	C4=0.5565E+00	K=0.2589E+01	GAIN=0.1610E+01
R5=0.1000E+01	C5=0.2345E+01	A=0.4264E+00	

N = 7	R_{db} = .18db	ω_s = 1.221	A_{db} = 55.42db
R1=0.3628E+00	C1=0.6063E+01	A=0.3281E+00	B=0.5978E+00
R2=0.7255E+00	C2=0.1347E+01	ZERO=0.2295E+01	
R3=0.2738E+01	C3=0.3570E+00		
R4=0.1232E+02	C4=0.1785E+00	K=0.1115E+01	GAIN=0.1025E+01
R1=0.4730E+00	C1=0.3439E+01	A=0.1493E+00	B=0.9100E+00
R2=0.9461E+00	C2=0.7641E+00	ZERO=0.1424E+01	
R3=0.7525E+00	C3=0.9607E+00		
R4=0.3386E+01	C4=0.4803E+00	K=0.2059E+01	GAIN=0.1451E+01
R1=0.5587E+00	C1=0.2637E+01	A=0.3905E-01	B=0.1017E+01
R2=0.1117E+01	C2=0.5860E+00	ZERO=0.1238E+01	
R3=0.5509E+00	C3=0.1188E+01		
R4=0.2479E+01	C4=0.5942E+00	K=0.2708E+01	GAIN=0.1616E+01
R5=0.1000E+01	C5=0.2238E+01	A=0.4469E+00	

N = 7	R_{db} = .18db	ω_s = 1.155	A_{db} = 48.81db
R1=0.3728E+00	C1=0.5632E+01	A=0.3309E+00	B=0.6332E+00
R2=0.7455E+00	C2=0.1252E+01	ZERO=0.2078E+01	
R3=0.2101E+01	C3=0.4440E+00		
R4=0.9456E+01	C4=0.2220E+00	K=0.1172E+01	GAIN=0.1049E+01
R1=0.4999E+00	C1=0.3207E+01	A=0.1354E+00	B=0.9258E+00
R2=0.9999E+00	C2=0.7126E+00	ZERO=0.1323E+01	
R3=0.6669E+00	C3=0.1068E+01		
R4=0.3001E+01	C4=0.5342E+00	K=0.2210E+01	GAIN=0.1474E+01
R1=0.5851E+00	C1=0.2524E+01	A=0.3292E-01	B=0.1015E+01
R2=0.1170E+01	C2=0.5608E+00	ZERO=0.1168E+01	
R3=0.5165E+00	C3=0.1271E+01		
R4=0.2324E+01	C4=0.6353E+00	K=0.2850E+01	GAIN=0.1623E+01
R5=0.1000E+01	C5=0.2092E+01	A=0.4781E+00	

$N = 7$	$R_{db} = .28db$	$\omega_s = 1.390$	$A_{db} = 70.19db$
R1=0.3493E+00	C1=0.7029E+01	A=0.2882E+00	B=0.5386E+00
R2=0.6986E+00	C2=0.1562E+01	ZERO=0.2789E+01	
R3=0.4856E+01	C3=0.2247E+00		
R4=0.2185E+02	C4=0.1123E+00	K=0.1087E+01	GAIN=0.1037E+01
R1=0.4277E+00	C1=0.3949E+01	A=0.1530E+00	B=0.8746E+00
R2=0.8555E+00	C2=0.8777E+00	ZERO=0.1668E+01	
R3=0.1006E+01	C3=0.7458E+00		
R4=0.4531E+01	C4=0.3729E+00	K=0.1844E+01	GAIN=0.1437E+01
R1=0.5047E+00	C1=0.2929E+01	A=0.4460E-01	B=0.1013E+01
R2=0.1009E+01	C2=0.6509E+00	ZERO=0.1415E+01	
R3=0.6544E+00	C3=0.1004E+01		
R4=0.2945E+01	C4=0.5021E+00	K=0.2435E+01	GAIN=0.1608E+01
R5=0.1000E+01	C5=0.2762E+01	A=0.3621E+00	

$N = 7$	$R_{db} = .28db$	$\omega_s = 1.325$	$A_{db} = 65.79db$
R1=0.3525E+00	C1=0.6819E+01	A=0.2904E+00	B=0.5524E+00
R2=0.7049E+00	C2=0.1515E+01	ZERO=0.2606E+01	
R3=0.4097E+01	C3=0.2607E+00		
R4=0.1844E+02	C4=0.1304E+00	K=0.1106E+01	GAIN=0.1046E+01
R1=0.4407E+00	C1=0.3803E+01	A=0.1478E+00	B=0.8825E+00
R2=0.8815E+00	C2=0.8452E+00	ZERO=0.1576E+01	
R3=0.9119E+00	C3=0.8170E+00		
R4=0.4103E+01	C4=0.4085E+00	K=0.1913E+01	GAIN=0.1446E+01
R1=0.5221E+00	C1=0.2834E+01	A=0.4175E-01	B=0.1012E+01
R2=0.1044E+01	C2=0.6297E+00	ZERO=0.1347E+01	
R3=0.6146E+00	C3=0.1070E+01		
R4=0.2765E+01	C4=0.5350E+00	K=0.2525E+01	GAIN=0.1612E+01
R5=0.1000E+01	C5=0.2684E+01	A=0.3726E+00	

N = 7	R_{db} = .28db	ω_s = 1.252	A_{db} = 60.18db
R1=0.3575E+00	C1=0.6518E+01	A=0.2932E+00	B=0.5731E+00
R2=0.7150E+00	C2=0.1448E+01	ZERO=0.2391E+01	
R3=0.3289E+01	C3=0.3149E+00		
R4=0.1480E+02	C4=0.1574E+00	K=0.1136E+01	GAIN=0.1059E+01
R1=0.4595E+00	C1=0.3608E+01	A=0.1399E+00	B=0.8939E+00
R2=0.9189E+00	C2=0.8018E+00	ZERO=0.1471E+01	
R3=0.8095E+00	C3=0.9102E+00		
R4=0.3643E+01	C4=0.4551E+00	K=0.2013E+01	GAIN=0.1461E+01
R1=0.5449E+00	C1=0.2718E+01	A=0.3774E-01	B=0.1012E+01
R2=0.1089E+01	C2=0.6040E+00	ZERO=0.1271E+01	
R3=0.5723E+00	C3=0.1150E+01		
R4=0.2575E+01	C4=0.5751E+00	K=0.2643E+01	GAIN=0.1617E+01
R5=0.1000E+01	C5=0.2571E+01	A=0.3889E+00	

N = 7	R_{db} = .28db	ω_s = 1.206	A_{db} = 56.12db
R1=0.3621E+00	C1=0.6272E+01	A=0.2952E+00	B=0.5908E+00
R2=0.7242E+00	C2=0.1394E+01	ZERO=0.2248E+01	
R3=0.2798E+01	C3=0.3608E+00		
R4=0.1259E+02	C4=0.1804E+00	K=0.1163E+01	GAIN=0.1070E+01
R1=0.4745E+00	C1=0.3463E+01	A=0.1333E+00	B=0.9030E+00
R2=0.9490E+00	C2=0.7696E+00	ZERO=0.1403E+01	
R3=0.7470E+00	C3=0.9777E+00		
R4=0.3362E+01	C4=0.4888E+00	K=0.2095E+01	GAIN=0.1472E+01
R1=0.5615E+00	C1=0.2640E+01	A=0.3458E-01	B=0.1011E+01
R2=0.1123E+01	C2=0.5866E+00	ZERO=0.1223E+01	
R3=0.5468E+00	C3=0.1204E+01		
R4=0.2461E+01	C4=0.6024E+00	K=0.2731E+01	GAIN=0.1621E+01
R5=0.1000E+01	C5=0.2481E+01	A=0.4030E+00	

N = 7	R_{db} = .28db	ω_s = 1.155	A_{db} = 50.86db
R1=0.3696E+00	C1=0.5919E+01	A=0.2974E+00	B=0.6177E+00
R2=0.7393E+00	C2=0.1315E+01	ZERO=0.2078E+01	
R3=0.2263E+01	C3=0.4297E+00		
R4=0.1018E+02	C4=0.2148E+00	K=0.1206E+01	GAIN=0.1088E+01
R1=0.4958E+00	C1=0.3274E+01	A=0.1234E+00	B=0.9157E+00
R2=0.9916E+00	C2=0.7276E+00	ZERO=0.1323E+01	
R3=0.6782E+00	C3=0.1063E+01		
R4=0.3052E+01	C4=0.5319E+00	K=0.2214E+01	GAIN=0.1489E+01
R1=0.5825E+00	C1=0.2548E+01	A=0.3018E-01	B=0.1010E+01
R2=0.1165E+01	C2=0.5661E+00	ZERO=0.1168E+01	
R3=0.5195E+00	C3=0.1270E+01		
R4=0.2338E+01	C4=0.6349E+00	K=0.2843E+01	GAIN=0.1627E+01
R5=0.1000E+01	C5=0.2352E+01	A=0.4252E+00	

TABLE 7.3a

Butterworth Normalized Polynomial Roots

Order N	Real Part $-\alpha$	Imaginary Part $\pm J\beta$
2	.7071	.7071
3	.5000	.8660
	1.0000	
4	.9239	.3827
	.3827	.9239
5	.8090	.5878
	.3090	.9511
	1.0000	
6	.9659	.2588
	.7071	.7071
	.2588	.9659
7	.9010	.4339
	.6235	.7818
	.2225	.9749
	1.0000	
8	.9808	.1951
	.8315	.5556
	.5556	.8315
	.1951	.9808
9	.9397	.3420
	.7660	.6428
	.5000	.8660
	.1737	.9848
	1.0000	
10	.9877	.1564
	.8910	.4540
	.7071	.7071
	.4540	.8910
	.1564	.9877

TABLE 7.3b

.1db Chebyshev Normalized Polynomial Roots

Order N	Real Part $-\alpha$	Imaginary Part $\pm J\beta$
2	.6125	.7124
3	.3500	.8695
	.6999	
4	.2183	.9262
	.5271	.3836
5	.3852	.5888
	.1471	.9527
	.4761	
6	.3926	.2591
	.2874	.7079
	.1052	.9671
7	.3186	.4343
	.2205	.7825
	.07869	.9758
	.3536	
8	.3065	.1953
	.2598	.5559
	.1736	.8320
	.06097	.9815
9	.2628	.3422
	.2142	.6432
	.1399	.8665
	.04857	.9854
	.2797	

TABLE 7.3c

.5db Chebyshev Normalized Polynomial Roots

Order N	Real Part $-\alpha$	Imaginary Part $\pm J\beta$
2	.5129	.7225
3	.2683	.8753
	.5366	
4	.3872	.3850
	.1605	.9297
5	.2767	.5902
	.1057	.9550
	.3420	
6	.2784	.2596
	.2037	.7091
	.07459	.9687
7	.2241	.4349
	.1550	.7836
	.05534	.9771
	.2487	
8	.2144	.1955
	.1817	.5565
	.1214	.8328
	.04264	.9824
9	.1831	.3425
	.1493	.6436
	.09743	.8671
	.03383	.9861
	.1949	

TABLE 7.3d

Maximally Flat (Bessel) Normalized Polynomial Roots

Order N	Real Part $-\alpha$	Imaginary Part $\pm J\beta$
2	1.1030	.6368
3	1.0509 1.3270	1.0025
4	1.3596 .9877	.4071 1.2476
5	1.3851 .9606 1.5069	.7201 1.4756
6	1.5735 1.3836 .9318	.3213 .9727 1.6640
7	1.6130 1.3797 .9104 1.6853	.5896 1.1923 1.8375
8	1.7627 .8955 1.3780 1.6419	.2737 2.0044 1.3926 .8253
9	1.8081 1.6532 1.3683 .8788 1.8575	.5126 1.0319 1.5685 2.1509

7.4 REVIEW OF OPERATIONAL AMPLIFIER FUNDAMENTALS

The operational amplifiers' versatility and low cost has made it one of the most popular building blocks in the industry. It can perform many mathematical processes upon signals. This section will briefly discuss some of the basic characteristics. An extensive formal analysis is covered in many standard texts and therefore will not be repeated here.

a. Characteristics of the Ideal Operational Amplifier

Figure 7-1 illustrates a simplified equivalent circuit for an operational amplifier. The output voltage e_o is the difference of the input voltages at the two input terminals amplified by gain A. A positive-going signal at the positive (+) input causes a positive change at the output whereas a positive-going signal applied to the negative (-) input causes a negative change at the output. The positive input is therefore called the "Non-Inverting" input and the negative input is called the "Inverting" input.

If the circuit of figure 7-1 were ideal, it would have the following parameters:

a. Infinite input resistance, R_i

b. Infinite gain, A

c. Zero ouput resistance, R_o

Negative feedback applied from the output to the inverting input results in zero differential input voltage. This basic property permits a wide variety of amplifier configurations and simplifies analysis.

An inverting amplifier is shown in figure 7-2. The differential input voltage is zero because of the negative feedback path through R_2. The inverting input is therefore at ground potential. The currents through R_1 and R_2 are:

$$I_1 = \frac{E_{in}}{R_1} \tag{7-1}$$

$$I_2 = -\frac{E_{out}}{R_2} \tag{7-2}$$

Since the amplifier input resistance is infinite, no current will flow into the amplifier. Therefore, $I_1 = I_2$. Equating equation 7-1 and 7-2 and then transposing results in the circuit gain expression:

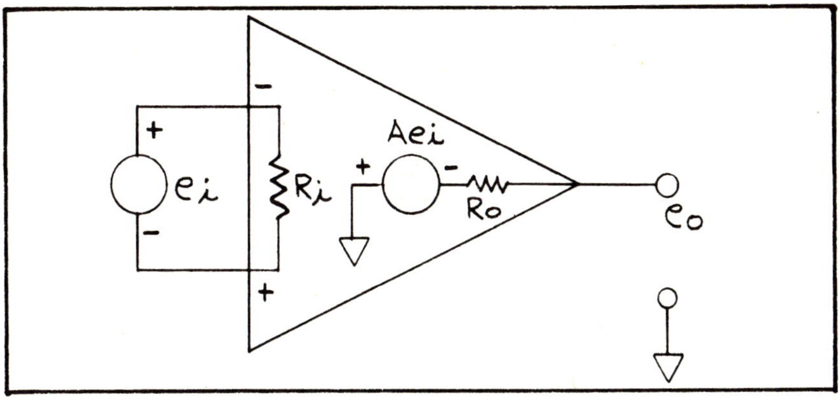

Figure 7-1
Simplified Equivalent Circuit of Operational Amplifier

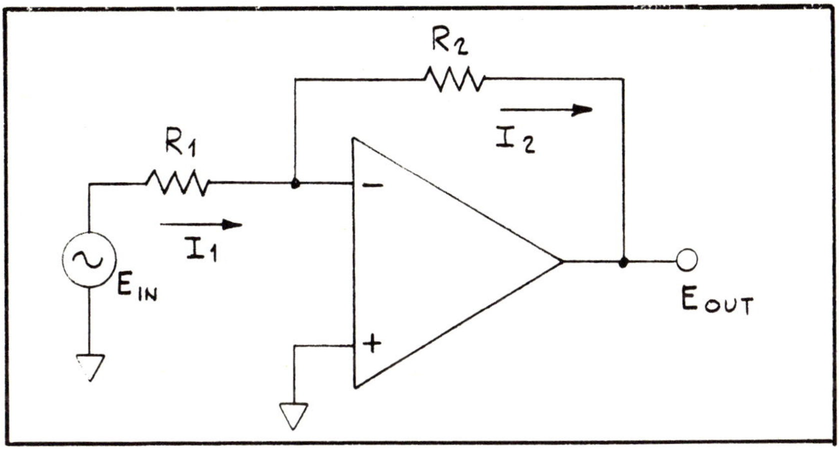

Figure 7-2
Inverting Amplifier

$$\frac{E_{out}}{E_{in}} = -\frac{R_2}{R_1} \quad (7\text{-}3)$$

The amplification is determined by the ratio of two resistors and is completely independent of the amplifier itself. The input resistance to the circuit is R_1 and the output resistance is zero.

The unique property at the inverting input where ground potential is maintained but no amplifier input current flows is called "Virtual

Ground". This effect leads to numerous applications. Two or more signals may be summed by connecting each signal through an associated resistor to the virtual ground. The current from each source is determined by its respective resistor value. These currents are algebraically combined and will flow through the feedback resistor producing a composite output signal. The three-input summing amplifier of figure 7-3 has the following output:

$$E_{out} = -\frac{R_2}{R_a} E_a - \frac{R_2}{R_b} E_b - \frac{R_2}{R_c} E_c \qquad (7\text{-}4)$$

The non-inverting amplifier of figure 7-4 can be analyzed in the following manner:

Since zero differential input voltage exists between the inverting and non-inverting amplifier inputs, the voltage across R_1 is equal to E_{in}. Since R_1 and R_2 form a voltage divider from E_{out}, and since the amp-amplifier draws no current, we can state:

$$\frac{E_{out}}{E_{in}} = \frac{R_1 + R_2}{R_1} \qquad (7\text{-}5)$$

or after simplifying:

$$\frac{E_{out}}{E_{in}} = 1 + \frac{R_2}{R_1} \qquad (7\text{-}6)$$

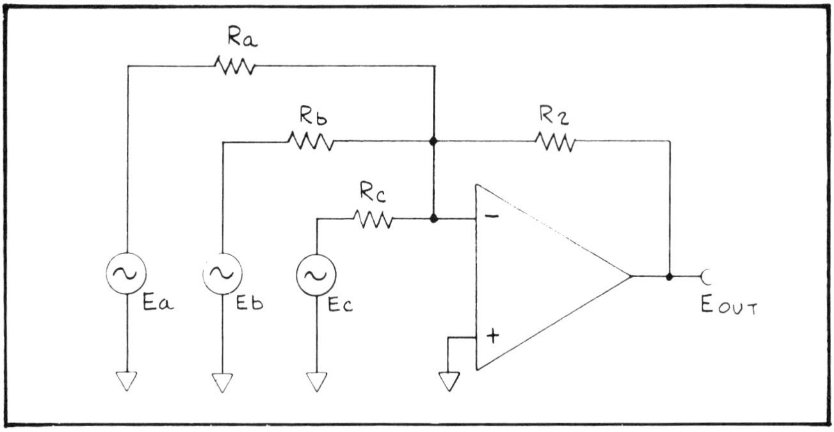

Figure 7-3
Summing Amplifier

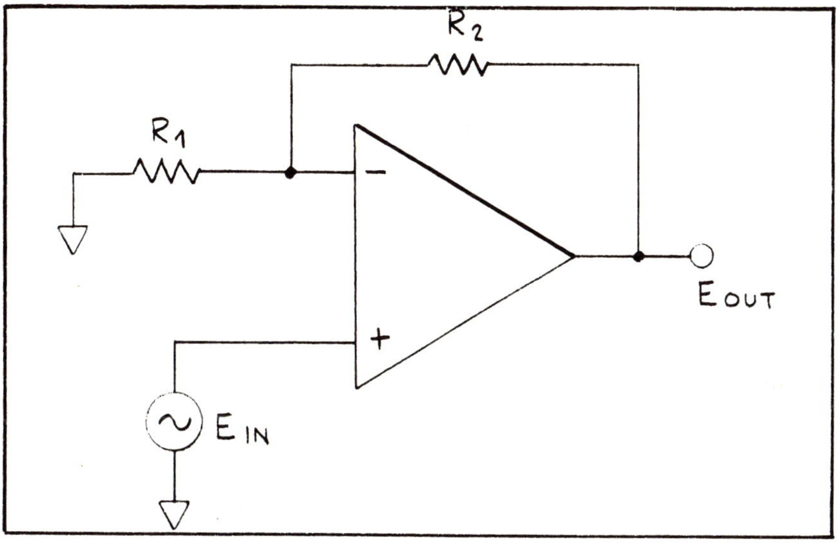

Figure 7-4
Non-Inverting Amplifier

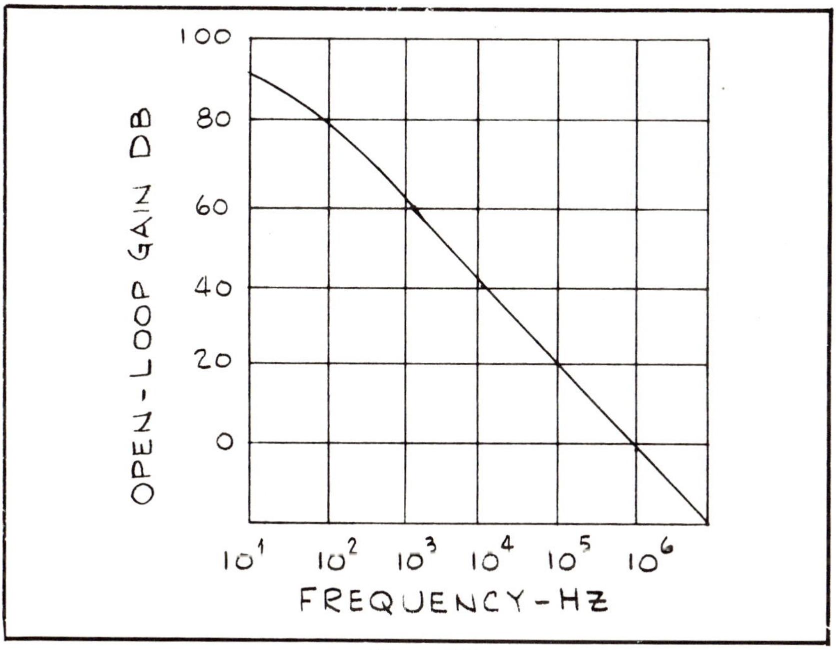

Figure 7-5
741 Open-Loop Gain VS Frequency

If R_2 is zero and R_1 is infinite, the circuit reduces to the Voltage Follower of figure 3-8a and the gain is unity.

Applications of operational amplifiers are by no means restricted to amplification and summing. If R_1 of figure 7-2 were replaced by a capacitor the circuit would function as a differentiator. A capacitor in in place of R_2 would produce an integrator. Non-Linear elements in the feedback loop may be used to perform Non-Linear operations.

b. Errors in Practical Amplifiers

The major sources of operational amplifier errors in active filters are gain-rolloff and DC offsets. Input and output impedance are usually of little consequence since filter network impedance levels can be chosen to minimize their effect (See Section 6.2b).

Gain Roll-Off

The gain of the basic amplifier without a feedback network is called Open-Loop gain. The gain with feedback is called Closed-Loop gain and was defined in Section 6.2a as:

$$A_c = \frac{A}{1 + A\beta} \tag{6-1}$$

where A is the open-loop gain and β is the feedback factor or fraction of the output applied to the input. If, at some frequency, $A\beta$ were equal to -1, equation 6-1 would blow-up to infinity which implies that the circuit would oscillate.

To prevent oscillation, the amplifier gain A must be band-limited so that $A\beta$ is less than 1 before the amplifier phase shift reaches 180°. This is accomplished by introducing a gain roll-off beginning at low frequencies and continuing at a 6db per octave rate. This approach to insure stability is called Frequency Compensation.

The feed-back factor of Figure 7-2 can be expressed as:

$$\beta = \frac{R_1}{R_1 + R_2} \tag{7-7}$$

It is clear that for high closed-loop gains, β diminishes. Less compensation can then be used which results in higher open-loop gain resulting in less gain error (see section 6.2a). The 709 operational amplifier provides this degree of freedom since compensation is externally connected. The 741 has built in compensation for the worst case possible; the voltage follower where $\beta = 1$. Figure 7-5 illustrates the

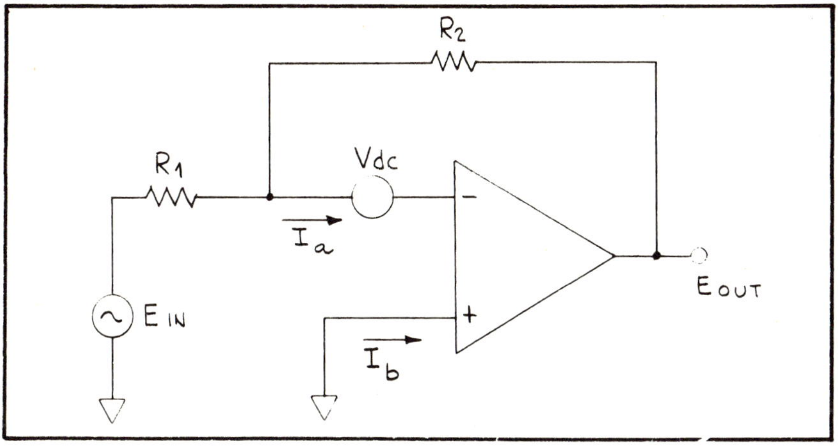

Figure 7-6
Inverting Amplifier with DC Offset

open-loop gain of this device. The response break point occurs very near to DC.

DC Offsets

The inverting amplifier of figure 7-2 is redrawn in figure 7-6 with the addition of the two sources of DC offsets, an offset voltage V_{dc} and two bias currents I_a and I_b. These effects occur within the amplifier because of circuit imperfections. Their magnitude will very from amplifier to amplifier and with temperature.

If we consider the effect of V_{dc} only, the output would be offset by an amount equal to V_{dc} multiplied by the closed-loop gain. The polarity of V_{dc} is random and the magnitude is typically less than 10 milli-volts.

The two bias currents I_a and I_b are nearly equal except for some offset (difference) current I_o. In the circuit of figure 7-6 I_b has no effect but I_a flows from R_1 and R_2 therefore producing an inverting input DC voltage offset equal to $I_a R_{eq}$ where

$$R_{eq} = \frac{R_1 R_2}{R_1 + R_2} \tag{7-8}$$

This voltage appears at the output multiplied by the closed-loop gain. It is clear that if R_{eq} were minimized the effect of the offset current I_a would be reduced. This is usually the most commonly used approach.

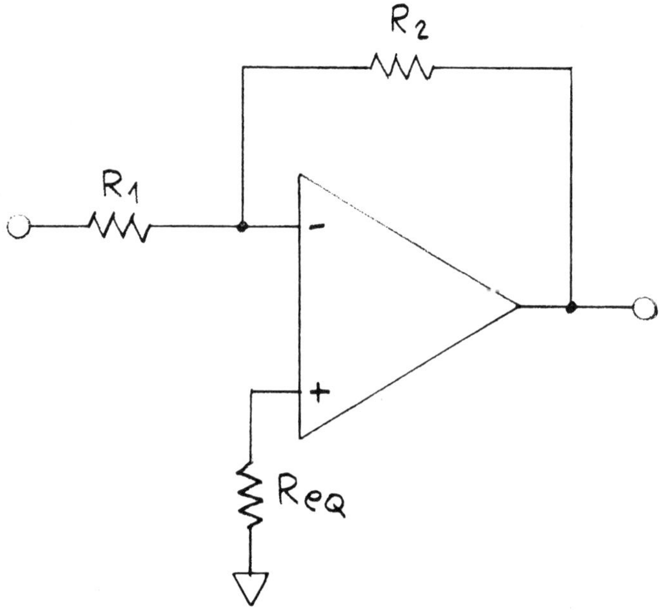

a. inverting amplifier

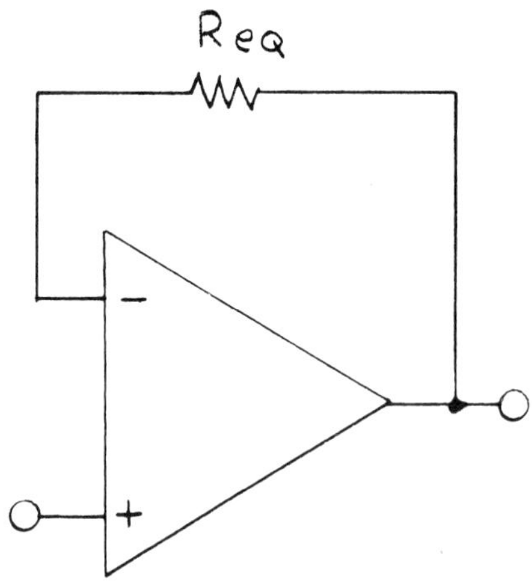

b. voltage follower

Figure 7-7
Minimizing Offset from Bias Current

An alternate approach can be used for more critical applications. If instead of returning the non-inverting input directly to ground a resistor having a value of R_{eq} can be introduced as in Figure 7-7a. This will have no effect on the overall circuit gain. However, a DC voltage is introduced at the non-inverting input equal to $I_b R_{eq}$ which is nearly equal to $I_a R_{eq}$, the offset at the inverting input. Since the amplifier is a differential device the DC voltage offset becomes $I_o R_{eq}$. I_a and I_b are typically 200 nano-amps and the offset current I_o is usually 30 nano-amps so this technique can offer a 7 to 1 reduction in the effect of bias currents.

In non-inverting amplifiers such as the one in figure 7-4 the ratio R_2 to R_1 is determined by gain formula 7-6. However, the actual values are nearly arbitrary and can be selected so that their parallel value (R_{eq}) is approximately equal to the total DC loading on the non-inverting input i.e. the parallel combination of all resistances connected between the non-inverting input and ground or other DC coupled sources. For a closed-loop gain A_c:

$$R_1 = \frac{A_c}{A_c - 1} R_{eq} \qquad (7\text{-}9)$$

$$R_2 = R_1 (A_c - 1) \qquad (7\text{-}10)$$

In the Voltage Follower of Figure 7-7b a feedback resistor of R_{eq} provides the necessary offset cancellation.

For moderate closed-loop gains or for AC coupled circuits the effects of DC offsets are usually of little consequence. Avoiding high impedance levels in the circuitry surrounding the amplifier is usually satisfactory.

For more critical applications such as precision active low-pass filters for recovery of small DC components the methods discussed can be implemented. Some amplifiers provide an input terminal for adjustment of output DC offsets.

c. Supply Voltage Consideration

Operational amplifiers operating in their linear region are relatively unaffected by DC supply voltage variations and power supply ripple. However, as a precaution against parasitic oscillation, the DC supply connections to each printed circuit card or amplifier should be decoupled to ground.

7.5 REFERENCES

1. Burr-Brown Research Corp. "Handbook of Operational Amplifier Active RC Networks", Tucson, Arizona: *Burr-Brown Research Corporation*, 1966.

2. E. Christian and E. Eisenmann "Filter Design Tables and Graphs", New York: *John Wiley*, 1966.

3. P. Geffe "Simplified Modern Filter Design" New York: *John F. Rider*, 1963.

4. J.N. Giles ED. "Linear Integrated Circuits Applications Handbook", Mountainview, California: *Fairchild Semiconductor*, 1967.

5. L.P. Huelsman, "Theory and Design of Active RC Circuits", New York: *McGraw-Hill*, 1968.

6. B.R. Shepard "Active Filters Part 12", *Electronics*, pp. 82-91, August 18, 1969.

7. L.C. Thomas, "The Biquad: Part I - Some Practical Design Considerations", *IEEE Trans. Circuit Theory*, Vol. CT-18, pp. 350-357, May 1971.

8. J. Tow, "A Step-by-Step Active-Filter Design", *IEEE Spectrum*, VOL 6, pp. 64-68, December, 1969.

9. M.K. Vander Kooi "Linear Applications", Santa Clara, California: *National Semiconductor Corporation*, 1973.

10. A.B. Williams, "An Active Equalizer with Adjustable Amplitude and Delay", *IEEE Trans. Circuit Theory*, Vol. CT-16, pp. 577-579, November 1969.

11. A.B. Williams, "Design Active Elliptic Filters Easily from Tables", *Electronic Design*, Vol. 19, No. 21, pp. 76-79, October 14, 1971.

12. A.I. Zverev, "Handbook of Filter Synthesis", New York: *John Wiley*, 1967.

Index

Active Filters
History, 1
Misconceptions, 2
Adjustable Delay and Amplitude
Equalizer, 93
Phase Shift Network, 100
Adjustment, 3
Band-Pass Filters, 83
All-Pass Delay Equalizer, 90
Delay Lines, 114
All-Pole Filters, 43
Band-Pass, 80
High-Pass, 48
Low-Pass, 44
Amplifier
DC Offsets, 183
Errors, 182
Gain Roll-off, 182
Ideal, 178
Impedance, 128
Non-Inverting, 67
Open Loop Gain, 126
Phase Shift, 126
Selection, 129
Summing, 180
Supply Voltage, 185
Voltage Follower, 43, 67, 184

Amplitude Equalizer, 93
Arithmetic Symmetry, 17

Band-Pass Filters
Alignment, 83
All-Pole, 80
Complex Pole Transformation, 81
Elliptic Function, 109
Group Delay, 20
Narrow Band, 15, 75
Real Pole Transformation, 80
Tolerances, 124
Wide Band, 15, 67
Band-Reject Filters, 19
Wide Band, 69
Boot Strapping, 89
Butterworth Attenuation, 22, 25
Group Delay, 26
Low-Pass Values, 138
Polynomial Roots, 172
Step Response, 27

Capacitor
Mica, 136
Mylar, 136
Poly Carbonate, 135
Polystyrene, 135
Selection, 135
Values, 137
Chebyshev Approximation, 24
Attenuation, 30, 33
Group Delay, 31,34
Low-Pass Values, 139—140
Polynomial Roots, 173—174
Step Response, 32, 35
Complex Frequency Plane, 9

Delay Equalizers, 90, 100
Delay Lines, 114
Dynamic Range, 129

Elliptic Function Filters
Band-Pass, 109
Comparison, 40
Delay, 40
High-Pass, 60, 106
Low-Pass, 56, 104
Response, 39, 49
State Variable, 100
Wide Band, 67

Frequency Response
Amplifiers, 180
Band-Pass Filters, 15
High-Pass Filters, 12
Low-Pass Filters, 12
Frequency Scaling, 3
Frequency Scaling Factor (FSF), 3, 44

Geometric Symmetry, 17
Group Delay, 19
of Band-Pass Filters, 20
of Low-Pass Filters, 20
Measurement, 119

High-Pass Filters
All Pole, 48
Elliptic Function, 49, 106
Frequency Response, 12

Impedance of Amplifier, 128
Scaling, 3

Lissajous Pattern, 88
Low-Pass Filters
All Pole, 44
Elliptic Function, 49, 104
Frequency Response, 12
Tolerances, 123

Maximally Flat Delay, 29
Attenuation, 36
Group Delay, 37
Low-Pass Values, 141
Polynomial Roots, 175
Step Response, 38
Mica Capacitors, 136
Mylar Capacitors, 136

Narrow Band-Pass Filters, 15, 75
Notch Frequency Tuning, 104
Notch Networks, 19, 89

Phase Shift Network, 100
Phase Shift of Amplifier, 126
Poles, 11
Complex Pole Band-Pass Transformation, 81
Real Pole Band-Pass Transformation, 80
Single Amplifier Pole-Pair Circuit, 75
State Variable Pole-Pair Circuit, 77
Polycarbonate Capacitors, 135
Polystyrene Capacitors, 135

Q
Adjustments, 88, 103
Amplifier, Effect of, 127
Definition, 89

Resistors, 131
Carbon, 131
Metal Film, 131
Standard Values, 134

Scaling, Frequency and Impedance, 3
State-Variable
Amplifier, Effect of, 124
Building Block, 101
Elliptic Function Filters, 100
Notch Adjustment, 104
Pole-Pair, 77
Q Adjustment, 103
Resonant Frequency Adjustment, 103
Steepness Factor, 14
Step Response, 22
of Band-Pass Filter, 22
of Low-Pass Filter, 22

Tolerance
of Band-Pass Filter, 124
of Low-Pass Filter, 123
Sensitivity, 123

Transfer Function, 9
Denormalizing, 11
Transient Response, 19
Twin T, 89

Voltage Follower, 43, 67, 184

Wide-Band
Band-Pass Filters, 15, 67
Band-Reject Filters, 69

Zeros, 11, 101